TELEPEN
80 8600550 9
DILEWYD O STOC
WITHDRAWN FROM STOCK
CPC UCW
LLYFRGELL
LIBRARY

AF598222

Cover:
A Lower Triassic ammonoid — *Otoceras boreale* Spath — from northern Axel Heiberg Island, slightly reduced in size.

THE TRIAS AND ITS AMMONOIDS:
THE EVOLUTION OF A TIME SCALE

E. T. Tozer

Geological Survey of Canada
Miscellaneous Report 35

1984

Available in Canada through

authorized bookstore agents
and other bookstores

or by mail from

Canadian Government Publishing Centre
Supply and Services Canada
Ottawa, Ontario, Canada K1A 0S9

and from

Geological Survey of Canada
601 Booth Street
Ottawa, Ontario, Canada K1A 0E8

A deposit copy of this publication is also available
for reference in public libraries across Canada

Cat. No. M41-8/35E — Canada: $8.00
ISBN 0-660-11600-6 — Other countries: $9.60
Price subject to change without notice

Preface

Much of the landmass of Canada exposes Phanerozoic sedimentary rocks, that is to say, fossiliferous strata of the Paleozoic, Mesozoic and Cenozoic eras. The study of these rocks is essential for elucidating the geological history of Canada and for assessing their economic potential. It is found that for some geological periods the record preserved in Canada contributes data of great significance towards understanding and interpreting the worldwide chronology. The most refined system of chronology in sedimentary rocks is biochronology, which is based on the interpretation of sequences of fossils.

Geological time scales of worldwide application benefit everyone. They open the way for compatible geological maps for different provinces, states and nations. Contemporaneity of events in earth history, some of which may be of great economic significance, can be better investigated. To achieve the best results international collaboration is essential; the fruits of such work are apparent in the pages that follow.

In recognition of the importance of defining time scales that can be used worldwide, the International Union of Geological Sciences (IUGS), through its Commission on Stratigraphy, has organized Subcommissions on each of the geological periods. The subcommissions, in turn, have established working groups to consider questions connected with the definition of period boundaries.

The Triassic period is one for which the Canadian rock sequences contribute particularly important data towards defining and recognizing the time scale. For this reason, Dr. Tozer, the authority on Triassic biochronology for the Geological Survey of Canada, was elected a Vice-Chairman of the Triassic Subcommission (STS) and Chairman of the working group on the Permian-Triassic boundary. He is also a member of the working group on the Triassic-Jurassic boundary. In these capacities he is required to make recommendations on the nomenclature and definition of boundaries, both at the limits and within the Triassic.

Préface

L'étude des roches sédimentaires du Phanérozoïque, dont les strates fossilifères datant du Paléozoïque, du Mésozoïque et du Cénozoïque affleurent sur une grande partie de la masse continentale du Canada, est essentielle pour bien comprendre l'histoire géologique du pays et pour en évaluer les possibilités économiques. Les successions conservées dans les formations canadiennes mises en place à certaines époques géologiques fournissent des données extrêmement utiles à la meilleur compréhension et interprétation de la géochronologie mondiale. La discipline la plus perfectionnée d'étude chronologique des roches sédimentaires est la biochronologie, qui repose sur l'interprétation des séquences de fossiles.

Les échelles de temps géologiques utilisées dans le monde entier servent des intérêts universels. Ainsi, elles ouvrent la voie au dressage de cartes geologiques compatibles des différentes provinces, territoires et nations. Elles permettent, en outre, de mieux étudier la contemporanéité de phénomènes géohistoriques, dont certains peuvent revêtir une grande importance économique. Pour obtenir les meilleurs résultats, la collaboration à l'échelle internationale est essentielle; on pourra d'ailleurs en constater les avantages manifestes dans le présent document.

Étant donné qu'il importait d'universaliser les échelles chronologiques, l'Union internationale des sciences géologiques (UISG), dans le cadre de sa commission stratigraphique, a établi des sous-commissions, chacune étant chargée d'une période géologique particulière. Ces sous-organismes ont à leur tour formé des groupes de travail pour étudier les questions liées à la définition des limites de ces périodes.

Le Trias est l'une des périodes pour lesquelles les séquences rocheuses du sol canadien ont produit des données particulièrement précieuses pour définir et mettre en application une échelle chronologique universelle. C'est pourquoi M. Tozer, reconnue autorité en matière de biochronolgie du Trias à la Commission géologique du Canada, a été élu vice-président de la sous-commission chargée du Trias et président du groupe de travail cherchant à délimiter le Permien du Trias. Par ailleurs, il fait aussi partie du groupe de travail cherchant à délimiter le Trias du Jurassique. Dans l'exercice de ces fonctions, il doit présenter des recommandations concernant la nomenclature et la définition des limites d'échelle, tant au haut et au bas du Trias qu'au sein même de cette période.

The gradual development of progressively more refined time scales has been going on for more than 150 years. For the Triassic, as with most of the geological periods, the activity started in western Europe. It was soon found that localities in other parts of the world, particularly the Himalayas, contributed data of as great, or even greater significance. Later it was found that North America also provided information of paramount importance, discoveries being made first in the United States and later on in Canada.

The nomenclature developed with these discoveries. Viewing the development is a fragment of scientific history, showing the extent to which we all stand on our predecessor's shoulders. But a historical approach is also important towards ensuring that terminology is correctly and uniformly applied – the aim of the Stratigraphy Subcommission. In detail it is a complicated subject. Difficulties arise, with a name being used in one country in one sense and elsewhere in another. The main job of the Stratigraphic Subcommissions is to identify and iron out these difficulties. Dr. Tozer has prepared this account as a contribution to the aims of the Triassic Subcommission, by summarizing the history, focusing on problems that need to be addressed, and making recommendations.

The report has not been submitted to the usual editorial procedures. Most of the literature research and writing was done in the author's own time, and he makes no apology for the intrusion of assorted historical side-lights and true tales of murder. He hopes that the historical aspects will interest readers outside the circle of Triassic specialists and for this reason has adopted a style which attempts to make a rather indigestible subject less unappetizing. The comprehensive index will enable the specialist to locate passages that deal with specific problems. An Appendix provides data on the nomenclature of Triassic chronology.

Ottawa, January 1984.

R.A.Price,
Director General,
Geological Survey of Canada.

L'élaboration graduelle d'échelles de temps de plus en plus précises se poursuit depuis plus de 150 ans. Pour le Trias, comme pour la plupart des autres périodes geologiques, les travaux ont été lancés en Europe de l'Ouest mais l'on a tôt fait de constater que des localités situées ailleurs dans le monde, surtout dans la région de l'Himalaya, renfermaient des données toutes aussi utiles, voire davantage. Plus tard, on a aussi pu apprécier la contribution de l'Amérique du Nord, suite aux découvertes faites d'abord aux États-Unis, puis au Canada.

La nomenclature s'est établie au gré de ces découvertes. L'étude de son évolution nous révèle une facette fragmentaire de la démarche scientifique, qui montre à quel point nous sommes tributaires des réalisations de nos prédécesseurs. Là ne s'arrête pas l'importance de l'étude historique dont la fonction est également d'assurer la mise au point d'une terminologie d'emploi correct et d'application uniforme, soit l'objectif de la sous-commission. Si l'on entre dans le détail, le sujet se complique. Des difficultés s'interposent, comme l'objet que désigne un nom qui varie d'un pays à un autre. La tâche première des sous-commissions stratigraphiques consiste donc à relever et à éliminer ces problèmes d'inégalité. M. Tozer a rédigé ce compte rendu, afin de contribuer a la réalisation des objectifs terminologiques de la sous-commission chargée du Trias, en résumant l'historique, en faisant le point sur les problèmes à résoudre et en suggérant des solutions.

La rapport ne respecte pas les normes habituelles de rédaction. La plupart des recherches documentaires et la plus grande partie de la rédaction ont été effectuées par l'auteur dans ses temps libres. En outre, il ne se sent nullement coupable d'avoir intercalé dans le rapport des digressions d'intérêt historique et des récits véridiques de meurtre. Il espère ainsi intéresser des lecteurs hors du cercle des spécialistes du Trias; il a donc adopté un style qui cherche a faciliter la lecture d'un sujet plutôt aride. Un index très complet permettra au spécialiste de retrouver les passages qui l'intéressent particulièrement, et un appendice fournit des explications sur la nomenclature utilisée pour désigner la chronologie du Trias.

Ottawa, janvier 1984.

Le directeur général
de la Commission géologique
du Canada,
R.A.Price.

Contents

Maps

Tables

Illustrations

I

Introduction

The Trias, or Triassic Period, the first of the Mesozoic Era, is now judged from radiometric data to have lasted for about 40 million years, starting about 245 and ending about 205 million years ago. The name Trias is derived from the three groups of rock, the Bunter, Muschelkalk and Keuper, which were formed during the period in western Europe. Oldest of the three is the Bunter. These rocks are best developed, and were first studied, in Germany. For this reason these typical Triassic rocks are referred to as the Germanic Trias.

Today marine Triassic rocks are known around the Pacific and Arctic oceans and in a belt extending from the Alps, east to Turkey, Iran, the Himalayas and Indonesia. This belt marks the site of the Triassic seaway known as Tethys. Marine Triassic is conspicuously absent on the Atlantic shores nor is it found in Atlantic drill holes. Generally speaking, in Triassic time the globe had one hemisphere mostly of land – Pangaea and another mostly of water – Panthalassa. In Panthalassa there may have been many island archipelagos.

Marine Triassic rocks contain abundant ammonoids. These are fossils of extinct shell-bearing cephalopods, with chambered shells, distantly related to the living *Nautilus*. Throughout their geological history of some 300 million years (Devonian to Cretaceous), ammonoids evolved rapidly and often explosively. Distinctive faunas appeared suddenly, achieved widespread, often worldwide distribution, disappeared, and were then replaced. At the end of the Trias they nearly became extinct. But it was not until the end of the Cretaceous that the ammonoids disappeared for ever.

The Triassic was the heyday of the ammonoids. During no other period was a comparable variety developed. About 55 successive faunas, many distributed worldwide, are recognized nowadays. These faunas provide the basis for a biochronological scale. Owing to their rapid rate of evolution ammonoids are in a class by themselves for biochronology. Nearly as useful in the Trias are the thin-shelled clams known as *Claraia, Enteropleura, Daonella, Halobia and Monotis*. These clams have no close counterparts in the living fauna. They were probably distant relatives of the one adopted as a trade mark by the well known oil company. Somehow species of *Monotis* etc. achieved worldwide distribution. It is most unlikely that they lived on the sea bottom, as do most living clams. Probably they lived attached to masses of seaweed and drifted to all parts of the Triassic seas. Because they are so important for Triassic biochronology, these clams get quite a bit of attention in the following pages.

Comparison with the radiometric data shows that the time span of each successive fauna was no more than a million years. Biochronology thus provides an immensely powerful and unmatched tool for deciphering earth history and establishing the contemporaneity of events in the distant past. Thanks to fossils, our knowledge of earth history since the Cambrian, i.e. over the past 600 million years or so, is much better than that of the Precambrian, which is still very imperfectly understood by comparison, owing to the

TABLE I. GEOLOGICAL TIME SCALE

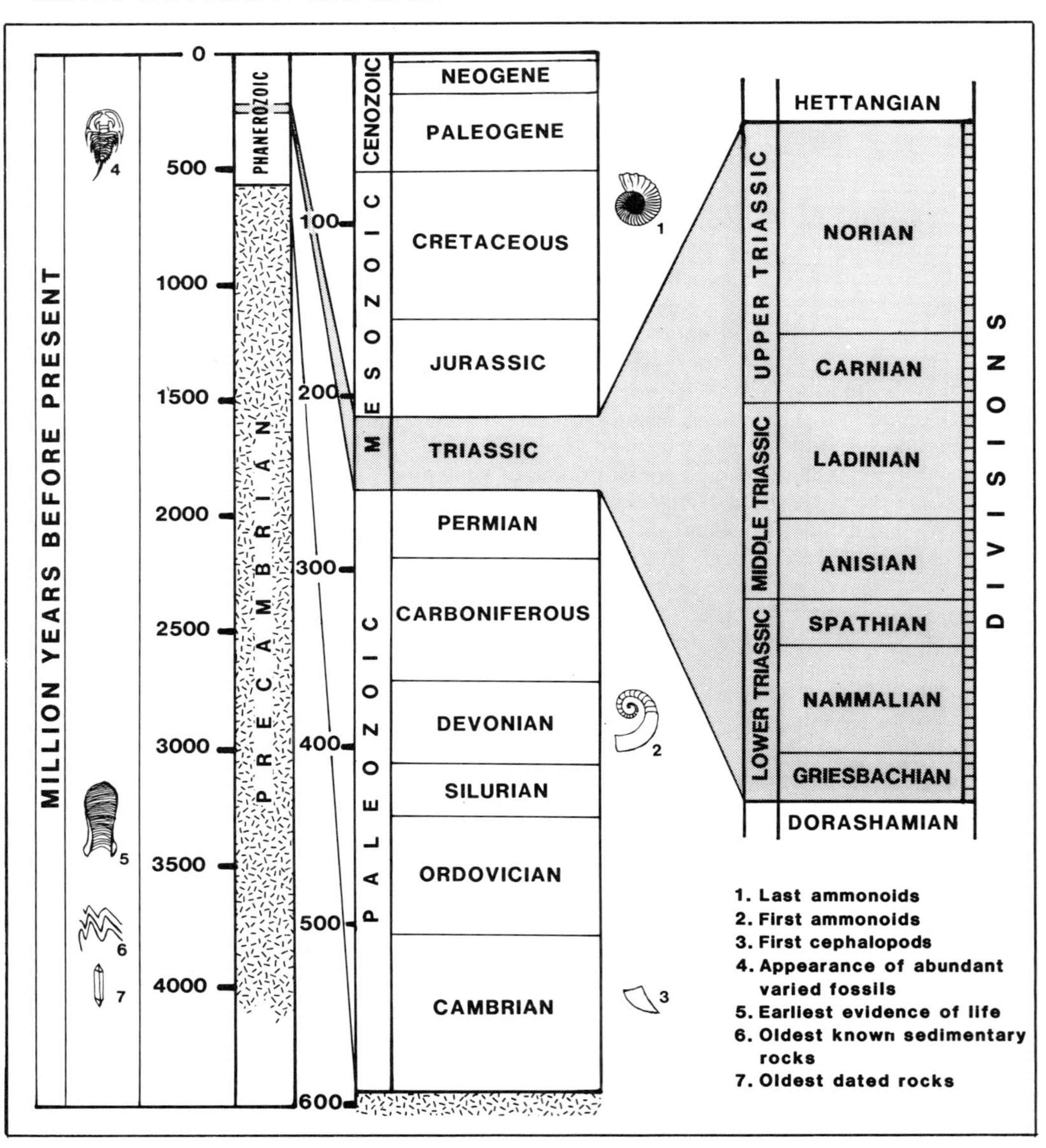

absence of biochronologically useful fossils. Biochronology is a field of research with many active participants; new data are still being acquired, and the time scale is continually being improved. This book deals with the history of its development for the Trias.

Establishment of a biochronological scale can only be achieved by assembling data on faunas observed in sequence in successive layers of sedimentary rock. This may seem a truism, but it needs emphasis because in the past, and even nowadays, attempts have been made to arrange faunas in sequence when they are not known in geological succession. Instead, they are placed in sequence on the grounds that one fauna is more "advanced", from an evolutionary standpoint, than another. This, of course, begs the question, for what is more advanced ? When dealing with ammonoids not known in geological succession, the question is unanswerable.

Because the successive ammonoid faunas changed so obviously and so frequently their potential for biochronology was soon recognized. For the same reason they attracted the attention of evolutionists. Studying the evolution of ammonoids when their succession in time is known from stratigraphy is a respectable scientific pursuit. Attempts to study their evolution from collections in museums, for which the sequence is not known, is dangerously unscientific. Although this may seem obvious it must be recorded, because it was the practice of old-timers such as Alpheus Hyatt, and there are some contemporary workers still doing this sort of thing.[1] Evolutionary studies of this sort led to the proposal of time scales for the Trias and other periods. Some became rooted in the literature, their shaky, indeed non-existent foundation forgotten or overlooked.

Time scales based solely on evolutionary ideas are always suspect. Many of the Triassic ammonoid localities found in the last century were isolated. Splendid collections were made, but the actual position of many a fossil bed in relation to others, simply could not be determined. With a fauna discovered on one mountain, and a different one in a nearby valley, inevitably, and quite reasonably the question would arise, which is the older of the two ? When the faunas were new, and known nowhere else in the world, there were no scientific grounds for deciding upon their relative age. But sometimes the temptation to make a decision was irresistable and time scales came into being for which there was no scientific foundation whatever. Real troubles arose when these time scales came to be accepted by people who did not have the time, nor feel the necessity, for examining the evidence on which they were based.

For 30 years I have been studying the marine Trias and its ammonoid faunas, mainly towards refining the biochronological scale by seeking faunas in sequence. The literature extends back to the late 18th century. Students of fossils, like zoologists, must pay attention to the old literature. Questions of priority must be considered, not only to do justice to one's predecessors, but in order to ensure stability and satisfy the International Rules of Zoological Nomenclature. The paleontologist thus inevitably becomes deeply involved in the history of his subject. More so, perhaps, than the ordinary geologist, or indeed most other scientists. Paleontological literature, except to the specialist, must be considered rather boring, although the beauty of the ammonoids, and the skill of the 19th century lithographers, resulted in illustrated monographs with aesthetic properties attractive to almost any bibliophile. The texts may have a limited appeal. It is easy to forget that the sometimes boring, tedious, detailed descriptions of the rocks and fossils were written by real people. This book attempts to focus on the people as well as the subject: their lives, friendships, alliances, jealousies and animosities, as well as their scientific contributions. Many, it turns out, were far from dull! I have undertaken this to satisfy my own curiosity but possibly the results will interest others besides the fraternity of Triassic specialists.

The tale developed in the following chapters is one of progressive acquisition of knowledge. Most chapters are identified with one or two dominant personalities. By putting

PANGAEA SIDE OF THE TRIASSIC GLOBE

Above: interpretation of the Triassic globe according to E. Irving.[258] Opposite: above, the Panthalassa side in Triassic time; below, the distribution of marine Triassic today.[259] Vertical pattern indicates areas of plate-bound marine Trias. In these areas the rocks are mostly sedimentary. They were deposited on the continental plates. Although in many places these rocks have been folded and faulted they have not moved extensively in relation to the plates since Triassic time. Horizontal pattern indicates marine Triassic of suspect terranes. Today these rocks are on terranes that may have moved hundreds or thousands of kilometres since the Trias, propelled by movement generated by sea floor spreading. These rocks are both sedimentary and volcanic. Dashed lines with arrows indicate the spreading ridges. The position of the ridge system in the Trias is a calculated guess. The patches numbered 1-25 were evidently islands, volcanic archipelagos, reefs, sea mounts and shoals. Since Triassic time 1-13 have been pushed east, amalgamating with North and South America; 17-25 have been pushed and squeezed in other directions by the movements that closed the Tethys seaway. 14, 15 and 16 show the probable position of northern New Guinea, New Caledonia and New Zealand in the Trias. Stars show occurrences of Hallstatt-type limestone; the cluster numbered 25 indicates occurrences now in the Mediterranean region; 24, the Kiogar mountains of Tibet; 23, Timor. Marine Triassic rocks are absent in the stippled areas. These areas were probably never covered by the Triassic seas. The marine reptiles are ichthyosaurs. They inhabited the Triassic seas in the company of the ammonoids. Ichthyosaurs were the Triassic counterparts of dolphins and whales. Some were about 15 metres long.

PANTHALASSA SIDE OF THE TRIASSIC GLOBE

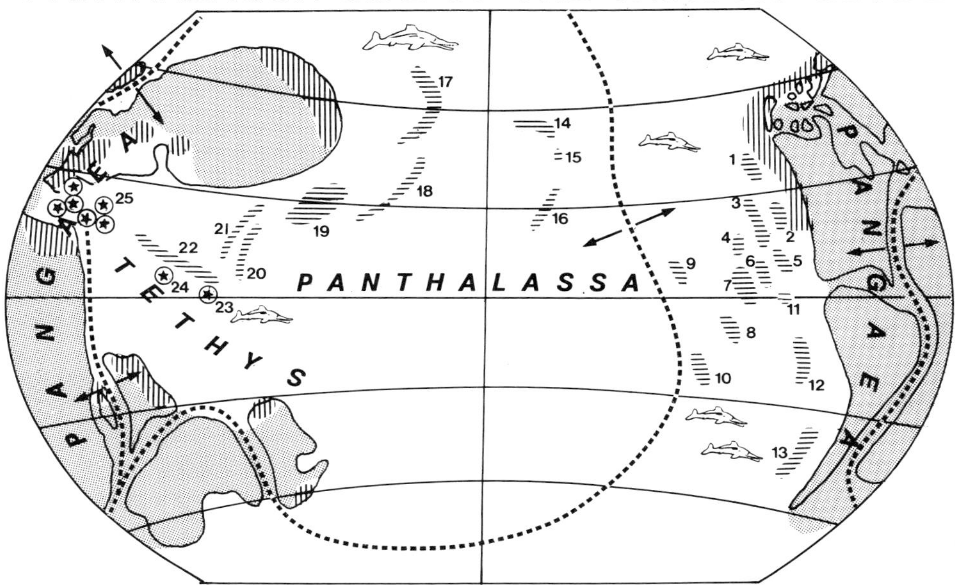

DISTRIBUTION OF MARINE TRIAS TODAY

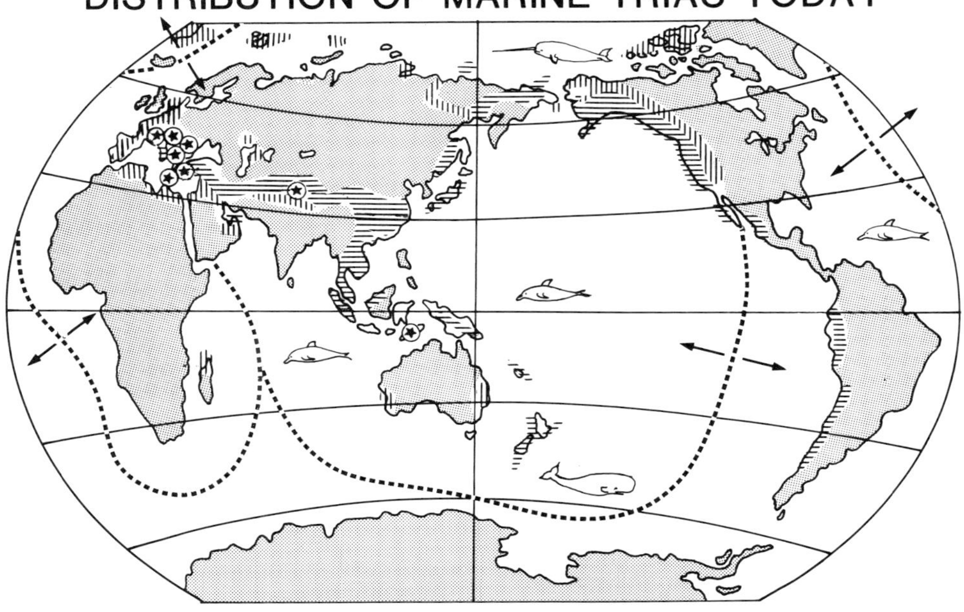

things in historical sequence and perspective it will be seen to what an enormous extent we all stand on our predecessors' shoulders. Much can be derived from reading the old publications.

The story begins in western Europe. Some foundations were laid as early as the 1500s, but most of the progress starts in middle of the 18th Century, from the work of Johann Gottlob Lehmann (died 1767) and Georg Christian Füchsel (1722-1773). Lehmann published his classic – "Versuch einer Geschichte des Flötzgebirge" in 1756. Füchsel's equally important book, "Historia terrae et maris, ex historia Thuringiae per montium descriptionem erecta", appeared in 1762. All of it is in Latin. Sir Archibald Geikie translates the title as "A History of the Earth and the Sea, based on a History of the Mountains of Thuringia".[2] I have not seen this rare book but it sounds like pretty good stuff ! Lehmann's Flötzgebirge is what today would be called sedimentary terrain. The term appears in Trias literature well into the 19th century. Lehmann and Füchsel seem to have been real pioneers by showing that the layers of the Flötzgebirge were amenable to scientific study and told tales about the history of the earth. It would seem that they invented stratigraphy, although some of their ideas may have been anticipated by Nikolaus Steno (1638-1687), who was born in Copenhagen and who travelled as a scholar throughout most of western Europe.[3] Lehmann and Füchsel recognized the Bunter (now the Lower Triassic). Füchsel actually named the Muschelkalk (Middle Trias). To put these people in historical perspective: their books were published well before those of the generally acknowledged pioneers, James Hutton (1726-1797), and William Smith (1769-1839). Hutton, a Scot, published his "Theory of the Earth" in 1785. Lehmann's and Füchsel's books were published before Smith, sometimes described as the father of stratigraphy, was even born. Lehmann, was an all-rounder, and also interested in chemistry. His scientific career ended abruptly in 1767 when he died from injuries caused by an explosion when he was experimenting with a retort filled with arsenic.[4]

In the 19th century, geology, including study of the Trias, entered its most exciting phase. Christian Leopold Freiherr von Buch (1774-1853), one of the geological giants of all time, had much to do with Trias in its early years.[5] Born in Germany, he was the son of a nobleman, evidently wealthy, because throughout his life von Buch was of independent means and never had to take a job. He had been a pupil of Werner's at Freiberg. Abraham Gottlob Werner (1749-1817) is generally acknowledged to have been a good mineralogist.[6] He also held strong views on earth history most of which were contrary to those of Lehmann, Füchsel, Hutton and Smith. Werner is now judged to have been wrong. His school was known as "Neptunist". He believed that crystalline rocks – granite and basalt – had been precipitated from a primeval ocean. Von Buch soon corrected some of the mistakes made by his teacher, for example proving that basalts were formed by volcanic action, not precipitates. A bachelor, energetic field geologist and brilliant linguist, he travelled all over Europe making geological observations and visiting his many friends, in both scientific and high social circles. His great stature as a geologist may be defined in unquestioned terms. But of his physical stature there are conflicting accounts. Sir Roderick Murchison described him as "The little vivacious man";[7] K.A. von Zittel, as "a tall imposing figure".[8] As the compiler of the first geological map of Germany von Buch knew the typical Trias well. He also worked in the Alps and made collections from the St. Cassian locality, which figures importantly later in the story.

By the mid-1800s Vienna was a focal point for all geology and particularly for Trias studies. It was the home of another giant: Eduard Suess (1831-1914).[9] Although Vienna became his home, where he was eventually a professor at the University, Suess was born in London, where his father, a Saxon, was a wool importer. Business in London lagged, however, and when Suess was three the family went to Vienna. He was educated there and in Prague. Besides writing the monumental "Face of the Earth", Suess was an active researcher on the Trias and ammonoids. Many of his students took up the work. He will appear several times in the story. Reading his works, and those of his contemporaries and students,I have become infected with the excitement that pervaded the geological community in Imperial Vienna of the 19th century. Much of the excitement, and, as it will be shown, much of the animosity, was concerned with the Trias, the leading personalities being von Hauer, Mojsisovics and Bittner.

Starting in 1810, but mainly from about mid-century, stray collections obtained by explorers in widely scattered parts of the earth – the Himalayas, the Arctic, New Zealand, California, British Columbia and Alaska, provided pieces that began to fit the puzzle.

Towards the end of the 19th century important new data were coming in from the Himalayas and the western United States. This leads to assessing the significant contributions of C.L. Griesbach, Carl Diener, A. von Krafft, James Perrin Smith, and several others.

The Great War of 1914-1918 brought the end of Imperial Vienna and heralded a period of doldrums in Triassic studies which lasted until the 1950s. In this period the most important things were being done elsewhere, notably by S.W. Muller in Nevada and F.H. McLearn in British Columbia. L.F. Spath, who wrote extensively on the Trias and its ammonoids, and who exercised great influence, is also of this period. Spath's influence is the more remarkable, because it was all based on study of collections in the British Museum. In his long career it would seem that not once did he collect a Triassic ammonoid, or examine an outcrop of marine Triassic rock. In the 1950s there was a renaissance. New data for refining the time scale came from all quarters; first from North America and the Arctic; later from the Tethys. Much of the Tethys activity has been achieved or spurred by scientists in Vienna. The ghost of Suess, I feel sure, must be deriving great satisfaction.

TABLE II. TRIASSIC TIME SCALE 1984

SERIES	STAGE	SUBSTAGE	ZONES (*STRATOTYPE IN TETHYS, OTHERS IN NORTH AMERICA)		
UPPER TRIASSIC	NORIAN	**Upper Norian**	*Choristoceras* ***crickmayi*** *Cochloceras* ***amoenum*** *Gnomohalorites* ***cordilleranus***		*Choristoceras* ***marshi******
		Middle Norian (Alaunian)	*Himavatites* ***columbianus*** *Drepanites* ***rutherfordi***		*Halorites* ***macer**** *Himavatites* ***hogarti**** *Cyrtopleurites* ***bicrenatus****
		Lower Norian	*Juvavites* ***magnus*** *Malayites* ***dawsoni*** *Stikinoceras* ***kerri***		*Malayites* ***paulckei**** *Guembelites* ***jandianus****
	CARNIAN	**Upper Carnian (Tuvalian)**	*Klamathites* ***macrolobatus*** *Tropites* ***welleri*** *Tropites* ***dilleri***		*Anatropites beds** *Tropites* ***subbullatus****
		Lower Carnian (Julian)	*Austrotrachyceras* ***obesum*** *Trachyceras* ***desatoyense***		*Austrotrachyceras* ***austriacum**** *Trachyceras* ***aonoides****
MIDDLE TRIASSIC	LADINIAN		*Frankites* ***sutherlandi*** *Maclearnoceras* ***maclearni*** *Meginoceras* ***meginae*** *Progonoceratites* ***poseidon*** *Eoprotrachyceras* ***subasperum***		*Protrachyceras* ***archelaus**** *Eoprotrachyceras* ***curionii****
	ANISIAN	**Upper Anisian (Illyrian)**	*Frechites* ***chischa*** *Frechites* ***deleeni***	*Frechites* ***occidentalis*** *Parafrechites* ***meeki*** *Gymnotoceras* ***rotelliformis***	*Ticinites* ***polymorphus**** *Paraceratites* ***trinodosus****
		Middle Anisian (Pelsonian)	*Anagymnotoceras* ***varium***	*Balatonites* ***shoshonensis*** *Acrochordiceras* ***hyatti***	"*Paraceratites*" ***binodosus**** *Anagymnotoceras* ***ismidicum**** *Nicomedites* ***osmani****
		Lower Anisian (Aegean)	*Lenotropites* ***caurus***		
LOWER TRIASSIC	SPATHIAN		*Keyserlingites* ***subrobustus*** "*Olenikites*" ***pilaticus***	*Neopopanoceras* ***haugi*** *Subcolumbites beds* *Columbites* ***parisianus***	*Tirolites* ***cassianus****
	NAMMALIAN	**Smithian**	*Wasatchites* ***tardus*** *Euflemingites* ***romunderi***		*Anasibirites* ***pluriformis**** *Hedenstroemia* ***himalayica****
		Dienerian	*Vavilovites* ***sverdrupi*** *Proptychites* ***candidus***		*Gyronites* ***frequens****
	GRIESBACHIAN	**Upper Griesbachian (Ellesmerian)**	*Proptychites* ***strigatus*** *Ophiceras* ***commune***		*Ophiceras* ***connectens****
		Lower Griesbachian (Gangetian)	*Otoceras* ***boreale*** *Otoceras* ***concavum***		*Otoceras* ***woodwardi****

II

Trias Divisions and Boundaries

From radiometric measurements it seems that the Trias lasted for about 40 million years (Ma). Biochronology makes possible the discrimination of about 55 divisions within the Trias. How are these divisions designated ? It might be supposed that it would be best not to have divisions, and instead speak of a Triassic rock or ammonoid as being 220 Ma old and so forth. This is impossible now, and probably always will be. Why ? Because the intercalibration between radiometric and biochronologic scales is still sliding with no immediate prospect of coming to grips. In 1964 it was estimated that the Triassic interval was from about 225 to 193 Ma; in 1978, about 247 to 212; in 1982, 245-204 Ma; later in 1982, 248-213 Ma; in 2000, who knows ?[10] Trias, however, seems to remain an interval of about 30-40 Ma. It is difficult enough to set the Triassic limits radiometrically. The prospect of providing accurate radiometric ages for the boundaries between the biochronological divisions is nil.[11] So although radiometry gives an idea of the length of time represented by the biochronological divisions there is still no prospect of designating them with radiometrically derived numbers. Biochronology requires its own system of terminology.

The terminology provides a hierarchy of divisions and subdivisions (Table II).[12] The parent division is the System – Trias. The geological terms "System" and "Period" are equivalent. A period is an interval of time; a system is the rock sequence on which the concept of the period is based. In practice they are more or less interchangeable. The biggest subdivisions of systems are series. Series are divided into stages. Some stages are divided into substages. Zones (standard zones) are the components of stages and substages. Zones, when amenable to division, are divided into subzones. The divisions down to substage level generally have geographic names; the smaller ones are named for fossils (e.g. *Frechites deleeni* Zone, or more concisely, Deleeni Zone). Despite the differences in nomenclature, all are divisions of the same kind. In order that they may be clearly defined it is absolutely necessary that all have a "stratotype", one place where a sequence of rock undoubtedly represents the division. Suggestions have been made that division boundaries should be defined in terms of more than one stratotype.[13] This procedure must be condemned because it can only result in ambiguity and confusion.

The hierarchy is flexible and accommodating. We shall see that in the early days only series were discriminated. Increase in knowledge led to the recognition of smaller divisions: stages and zones. The smaller divisions are grouped, sometimes of necessity arbitrarily, in the larger. As the biochronology becomes increasingly refined, with the recognition of progressively smaller divisions, the stratotype of the smallest division assumes the greatest importance. This is how the hierarchy works. A stage, until subdivided, can be interpreted only by its stratotype. When divided, for practical purposes it becomes defined in terms of its components. For stability and flexibility it is necessary that only the lower boundary of any division be defined. The base of zone A is defined by a bedding plane at a specified locality; the base of the next, B, is similarly defined. All the rock deposited in the time that elapsed between the formation of basal A and basal B is representative of zone A. Zones are discriminated, distinguished and recognized elsewhere by their distinctive faunas. The index fossil serves as a label, but it is the whole fauna that counts. Zones defined in this way may be compared with the reigns of monarchs. Provided that the dynasty is unbroken, historical events can be objectively and unambiguously expressed in terms of the reign. The statement that Columbus sailed to the Americas in the reign of the English King Henry VII is no less objective and unambiguous than saying that he did so in 1492.

It should not be supposed that the sequence of rock at the stratotype of zone "A" provides a complete sedimentary record for the time. It would be nice if it did, but it is becoming increasingly apparent that complete sedimentary records are the exception rather than the rule, even in the cores from drill holes in the deep ocean basins. In selecting a stratotype one hopes to approach the ideal of continuity but it is seldom, if ever, attainable.

It might also be supposed that the establishment of a biochronological time scale is easy, simply involving going to one place and driving in spikes to mark all the Triassic boundaries and thereby settling the matter once and for all. But nature has made things difficult. For the Triassic and the other geological systems, there is no such place. Or at any rate, if there is a place with sediment representing much of the Trias, all stacked up, the rocks do not contain an adequate representation of ammonoid faunas to provide a standard for comparisons. Of the 55 or so divisions within the Trias, each characterized by distinctive ammonoid faunas, nature has contrived that seldom can more than a dozen be seen in sequence in one place. The whole scale has to be built up by co-ordinating and correlating data derived from different localities. The bases of zones A, B and C may be defined at locality X; zones D, E and F at locality Y. The interpretation of zone D as being younger than zone C is not objective. It is the responsibility of the investigator to provide evidence that zone D overlies a correlative of zone C. This is not always easy, and is sometimes a matter of dispute. It is because of these difficulties and disputes that the establishment of the time scale is a continuing field for research.

Provided that the simple hierarchial rules are obeyed the science can advance in an orderly fashion. In the past, and even today, these rules have not always been followed, and as a result there have been, and are still, a lot of arguments about nomenclature. Some of the divisions proposed in the early days were framed in such vague terms that they have proved almost impossible to use. When names like this have been relegated to limbo, and have been replaced by new names with clear definitions, the old ones should be left in limbo. When they have been revived or revitalized, the judgement of the first reviser should be accepted, provided that it is legally defensible. This problem arises with the basal Triassic stage. In 1965 I proposed Griesbachian for this division. I recognized that Gangetian, proposed in 1895, certainly covered the Lower Griesbachian, less certainly the Upper Griesbachian. It was recommended that Gangetian be ranked as a substage, synonymous with Lower Griesbachian. To use it thus is certainly within the spirit of the original definition. It was the judgement of the first reviser. Now it has been suggested that Gangetian and Griesbachian should now be treated as absolute synonyms. This is deplorable. For 70 years Gangetian lay dormant. It was then revived, with one, precise, meaning: Lower Griesbachian. It should stay that way.[14]

Substages with geographic names are more popular in Europe than North America. In North America we tend to designate substages with a prefix, e.g. Lower Norian. Some Europeans prefer to call this interval "Lac", or "Lacian". This is an example of illegal resurrection. Lac was proposed by Mojsisovics in 1895 with a clearly defined stratotype.[15] In 1895 he also defined another substage – Sevat. He thought Lac was much older than Sevat but it was eventually found that they included beds of the same age. In 1902 Mojsisovics added some more beds to the Lac. These beds are now known to be older than Sevat, but in no sense can they be regarded as having been part of the original Lac.[16] I judge application of the name Lac to the Lower Norian illegal. I take this stand on legal grounds, but also to support a claim that I am capable of framing at least one sentence in the German language. In 1966 I was at the Geologischen Bundesanstalt in Vienna. Their building was the Rasumofsky Palace in the old days. I was looking at Mojsisovics' type specimens in a delightful room, overlooking the garden. Here, I am told, Beethoven once played the piano. Two other visitors arrived: Dimitrij Andrusov, the doyen of Carpathian geologists, who regrettably died a few years ago, accompanied by his wife, Vanda Kollarova-Andrusovova. They had driven over from their home in Bratislava, where Vanda does research on Triassic ammonoids. We had not met before, but because we were all looking at Triassic ammonoids, clearly we had common interests. We made efforts to discuss Trias problems. There was, however, some difficulty with language. Dimitrij and Vanda spoke German and Slovak; Vanda some French as well. They had very little English. I decided to try my German. I had recently published a paper in which Mojsisovics' chronology for the Norian was interpreted as being wrong.[17] In spite of the communication problem it was clear that they were genuinely interested in what I thought. I struggled to find words to express my interpretation and came up with "Lac ist Sevat". The message got through. They were surprised but agreed that I had a good case. We went off to lunch together and became good friends. But one should not get too exercised about these legal problems. I hope that none in the Triassic fraternity will try to generate strife of the kind that developed between Mojsisovics and Bittner in the 1890s (Chapter VIII). Bittner's first salvo was entitled "Was ist norisch ?"; his last was "Was ist lacisch ?". Even with my German I could probably assemble a new version of "Was ist lacisch ?". I would only have to select sentences from Bittner's voluminous writings, transposing lacisch for norisch where necessary !

Table II is a more or less up-to-date biochronological table for the Trias.[18] At the zone level it is based on the North American sequences. This is the only one with documentation of precisely defined zones covering the whole Trias. Since the establishment of the sequence in North America many of the Tethys zones have found their proper place. It is now possible to indicate Tethys zones more or less correlative with those defined in North America. They are also shown in the Table. Most of the zones are amenable to subzonal division. At present 58 divisions are recognized in the Trias of North America. On this subject there are still many unpublished data, consequently not all the divisions appear on Table II.

The names of the Middle and Upper Triassic stages date from the last century and are based on stratotypes in the Alps. For the Lower Trias, the Nammalian stratotype is in the Salt Range; Gangetian in the Himalayas; the remainder are in the Canadian Arctic, on Svartfjeld Peninsula, Ellesmere Island and Griesbach Creek, Axel Heiberg Island. Russian workers generally have two Lower Triassic stages: Induan, followed by Olenekian. The boundary roughly corresponds to that between the Dienerian and Smithian. Induan and Olenekian have not proved very suitable for worldwide application owing to ambiguity about the position of their mutual boundary. Some Russian workers are now using the Lower Trias divisions shown on Table II.[19] Russian workers have also introduced some new names for parts of the Lower and Middle Triassic.[20]

In Japan and New Zealand alternative stage names have been proposed for the whole Trias.[21] They are not used anywhere else.

There is another form of biochronological nomenclature, the kind that was developed by Sidney Savory Buckman (1860-1929). Buckman was an Englishman who worked extensively on Jurassic ammonoids and chronology. He also wrote on other subjects, e.g. "Mating, Marriage and the Status of Woman" under the pen-name James Corin.[22] Buckman was okay ! The book is anthropological, not male chauvinist.

Buckman's writings on biochronology appeared between 1891 and 1925. His early works are in scientific journals. Later there were troubles with editors and most of his ideas were propagated in a publication of his own. This was "Type Ammonites" (originally "Yorkshire Type Ammonites") which he started in 1909. The final part was published posthumously in 1930. In Buckman's schemes all biochronological divisions in the hierarchy were defined solely in biological terms and had names purely of biological connotation. A stratotype played no role in the definition of his divisions. Buckman was determined to make the distinction between the rocks and the time during which they were deposited. It may seem pretty obvious nowadays, but the geological terminology when Buckman arrived on the scene did not make the distinction clear. Triassic, Jurassic and Cretaceous, being defined by rocks, should not be used for periods of time. In 1922 Buckman replaced them with the Ceratitoidic, Ammonitoidic and Baculitoidic Periods. These derived their names from what Buckman considered to be the dominant ammonoids of the period. Buckman divided the periods into epochs, epochs into ages, ages into hemerae. All had names of biological derivation. In this way the time of *Psiloceras*, at the beginning of the Ammonitoidic ('Jurassic'), became the Psiloceratan Age, the first of the 47 that he eventually recognized. In 1898, before he had suggested Ceratitoidic and Ammonitoidic, he pointedly wrote of "so-called 'Jurassic' time", and generally placed both Jurassic and Triassic in quotation marks. So he was philosophically consistent from the beginning. In 1898 he also suggested that the natural faunal boundary between the 'Triassic' and 'Jurassic' took place later in time than had been supposed by other workers. Expressed in contemporary terminology, Buckman regarded the Hettangian ammonoids as Triassic rather than Jurassic. Hettangian faunas had been regarded as Jurassic by Oppel, before Buckman's time. They have been similarly assessed by virtually everybody else. Buckman did not hold this view very strongly, and eventually abandoned it himself.

L.F. Spath (Chapter XIII) adopted Buckman's principles and divided the Trias into 16 ages. He did not use "Ceratitoidic" in place of Trias. Nobody else has either. Provided that the concepts are based on observed sequences, chronologies like Buckman's have much to recommend them. But both Buckman and Spath got carried away and produced chronologies based to a considerable extent on supposed, not observed relationships. Their schemes are now pretty well discredited. The idea nevertheless has merit. Buckman's ages, although not defined with the precision of standard zones, nevertheless provide a terminology for expressing features of ammonoid evolution. If zones be compared with reigns, beginning and ending at a precise time, ages are comparable with concepts like the Bronze age and the Industrial Revolution. Not very precisely defined, perhaps, but nevertheless useful, significant, historical concepts.

Buckman should never be dismissed as a crank. He was certainly one of the first to express well reasoned thoughts on matters of biochronology. His contemporary, the Yale professor Henry Shaler Williams (1847-1918), was another. As a young man Buckman was an energetic field worker and made important contributions. In later years he resorted to interpreting the chronology from the fossils alone, without sequential data, and this led him to excesses for which he has been justly criticized.

Biochronological divisions with stratotypes, like those of Table II, are best. Do some of these divisions reflect dramatic events, or sudden changes, in the history of the earth ? It would be good to know. Some almost certainly do, although the nature of the event or change is still largely a matter for speculation. The boundary between the Permian and Trias, as defined by the base of the *Otoceras woodwardi* Zone, certainly seems to mark an interval when, in the words of D.J.McLaren, "something happened".[23] There is a big difference between the faunas above and below. In the marine sequences, a distinct bedding plane always seems to mark the boundary. Bedding planes admittedly can be highly ambiguous. Sometimes a distinct plane seems, from the biochronology, to indicate little change, and no great passage of time. Elsewhere, biochronology can demonstrate, within a metre or so of rock, in which there is no obvious lithological indication of interruption, that several million years of history have left no record. But the Permo-Triassic boundary bedding plane does seem to be a significant worldwide feature. But of what significance, we don't really know. There are still many unsolved mysteries in the stratigraphical record.

The boundary between the Triassic and Jurassic also marks a notable event in the history of marine life.[24] Ammonoids were at peak variety in the Middle Norian. By Upper Norian time many had gone, and the small Triassic heteromorphs inherited the seas, in the company of smooth big fellows: *Pinacoceras* and *Arcestes*. Also around were the Triassic ancestors of Jurassic *Phylloceras*. *Phylloceras* descendants occur in the earliest Jurassic rocks, but all the others had gone. Ammonoids undoubtedly nearly became extinct between latest Triassic (Norian) and earliest Jurassic (Hettangian). The record is mostly clearly displayed in Nevada (Chapter XIV). Here rocks with the latest Triassic ammonoids are followed by beds with those of earliest Jurassic age. Despite the change in the fauna there is little or no evidence for an interruption in sedimentation. Something catastrophic nevertheless seems to have happened. An explanation has been suggested. According to radiometric age determinations, about 210 Ma ago the rocks of the Manicouagan crater in Quebec were molten. This crater is a giant – about 70 km in diameter. Many (but not all) scientists believe that the crater was formed by the impact of a meteorite and that the melting was the result of the impact. The worldwide consequences of such an event would have been dramatic and disastrous as pointed out by D.J.McLaren, the "something happened" man. He has also pointed out that the time when the rocks were molten, from the radiometry, is close to the time given for the Triassic-Jurassic boundary.[25] So perhaps that's what did happen.

Discussion of the Triassic-Jurassic boundary has been hampered by problems of correlation and nomenclature. This has to do with the Rhaetian Stage.[26] This was named for the uppermost Trias in 1861 by the Munich professor, Carl Wilhelm Ritter von Gümbel (1823-1898). The stratotype – the Kössen Beds – are in the northern Alps. Quite a bit was known about the Kössen Beds before Gümbel named the Rhaetian, particularly from the work of Oppel and Suess.

Suess we have already met (Chapter I). Albert Oppel (1831-1865), already mentioned in connection with Buckman's views on the beginning of Jurassic time, must be ranked as the inventor of refined biochronology. Oppel studied in Tübingen with the renowned Jurassic worker, Friedrich August Quenstedt (1809-1889) and later became a professor at Munich. By 1856, when only 25 years old, he had produced a biochronological division of the Jurassic with 33 zones. Sad to say he died of typhoid at the early age of 34.

In 1856 Oppel and Suess gave evidence for correlating the Alpine Kössen Beds with the formation that Alberti, in 1834, classed as uppermost Triassic – the Täbingen Sandstone (Chapter IV). A fossil common to the Kössen and Täbingen formations is *Avicula contorta*. From these occurrences the name "Contorta Zone" came into use. Early in the 19th Century the fauna of the Contorta Zone was found to be widespread in northwest Europe, lying between red Keuper beds and the Lias, i.e. between unquestioned Trias and unquestioned Jurassic. Whether or not the Oppel-Suess correlation is exactly correct remains a problem to this day. But it is at least approximately correct. Its recognition provided an important step towards relating the Mesozoic stratigraphy of the Alps with that of the Extra-Alpine area.

After the publication of Gümbel's results, it became known, in 1865, from Hauer's work, that the fauna of the stratotype Rhaetian included a distinctive ammonoid – *Choristoceras* (Chapter V). In 1876 Suess and Mojsisovics showed that *Choristoceras* occurs above *Avicula contorta*. By 1895, when it was reviewed by the German professor, Josef Felix Pompeckj (1867-1930), the Rhaetian ammonoid fauna was fairly well known. In 1972 Max Urlichs made the important discovery that the ammonoid fauna of the Kössen Beds also included *Rhabdoceras* at a level just below *Choristoceras*. *Choristoceras* and *Rhabdoceras* also occur in the Norian. In British Columbia they occur together. This leaves no doubt that the Rhaetian ammonoid fauna has clear Triassic affinities. Before this was known, around the middle of the 19th Century, many scientists chose to class the Rhaetian and the Contorta Zone with the Jurassic. A. von Dittmar (Chapter VIII) summarized the situation in 1864, finding 18 in favour of a Triassic age, 12 undecided, and 21 for the Jurassic. If nothing else, this seems to show that a lot of people were interested in the question ! Hauer, in his book on the geology of the Austro-Hungarian Monarchy, published in 1875, had the Rhaetian between the Triassic and Jurassic, ranked as an equal (i.e. as a system), but this procedure did not catch on. Unfortunately the Extra-Alpine Rhaetian beds have no ammonoids and this is why the exact correlation with the Alpine sequence remains uncertain. In recent years there seemed to be evidence emerging that fossil spores confirmed the correlation advocated by Oppel and Suess. But the most recent conclusion seems to be that spores cannot be used for precise correlation at this level.[27]

Nowadays nearly everybody regards the Rhaetian as Triassic. There remains the question of its status. Stratotype Rhaetian is certainly followed by lowermost Jurassic – Hettangian. Problematic, however, is the age of the lower boundary of the Rhaetian and the exact relationship with the Norian, which traditionally was regarded as the stage preceding the Rhaetian. Norian, like Rhaetian, was first named in the northern Alps, but not at the same place. Norian as now understood, has a satisfactory base, but in the type area the succession up into the Jurassic is not clearly displayed. If it is, it has not been described. This means that there is no satisfactory boundary between stratotype Rhaetian and Norian. All the Rhaetian ammonoids, including *Choristoceras*, also occur in the Norian. The scope of the Norian, as it has been used throughout the 20th century, almost certainly overlaps that of stratotype Rhaetian. The scope of the Rhaetian is small, not comparable with the other Triassic divisions of stage rank. It is not demonstrably more than one zone (Crickmayi, Table II). In view of their probable overlap, Upper Norian and Rhaetian are best treated as synonymous substages. Rhaetian has historical priority over Norian but it seems undesirable to apply the letter of the law, and suppress Norian.

The stratotypic earliest Jurassic rocks, those of the Hettangian Stage characterized by *Psiloceras planorbis*, are Extra-Alpine. Below are beds of the Contorta Zone. Hettangian ammonoids also occur in the Alps, above the stratotype Rhaetian. Because stratotype Rhaetian is Alpine, and stratotype basal Jurassic Extra-Alpine, it is difficult to provide a satisfactory definition for the Triassic-Jurassic boundary in Europe. Fortunately there are illuminating sequences in North America for this interval (Chapter XIV). It is possible that some of the ammonoids that overly the Alpine Rhaetian and its equivalents in North America are a little older than the oldest in the stratotype Hettangian. This remains a problem. If true, on legal grounds the Alpine and North American ones would be Triassic ! That I think one would say, would be a case of the tail wagging the dog. Laws are necessary for stability, but should never produce biochronological divisions that obscure world-wide events in the earth's history.

III

Ammonoids and their Names

Ammonoids get their name from the Egyptian god Ammon (= Amen or Amun),[28] who became the supreme deity with the victory of the XVIII Dynasty over the Hyksos, around 1570 BC. The rulers of this dynasty, which lasted until 1342 BC, included a succession of four Amenhoteps. Ammon was usually represented in human form with a ram's head. It was the resemblance between the ram's horns and the corrugated spiral form of the fossils that led to the adoption of the name.

Because they are found in rocks all over the world, and are attractive, interesting objects, ammonoids were discovered by man as long ago as the late Paleolithic. In 1546 Georg Bauer (1494-1555), whose pen name was Agricola, published "De Natura Fossilium".[29] His activities centred around Chemnitz, now Karl-Marx-Stadt, East Germany. In this book ammonoids were a mineral, known as *Ammonis cornu*. Agricola did not provide any illustrations. But he describes a "golden armatura" on specimens from Hildesheim (south of Hannover). This means that they were preserved in marcasite or pyrite. Preserved in this way, and coming from Hildesheim, his specimens were almost certainly Jurassic, not Triassic ammonoids.

"De Rerum Fossilium", published in 1565 by the Zurich scholar Konrad Gesner (1516-1565) is probably the first book with illustrations of ammonoids.[30] Gesner seems to have been uncertain as to whether they were organic or inorganic.

The brilliant and versatile Robert Hooke (1635-1703) should be credited with the first accurate observations on the morphology of ammonoids and for interpreting them as the remains of extinct cephalopods. Hooke, who was born on the Isle of Wight, has been justly described as one of the founders of modern geology. He was connected with the Royal Society of London from its earliest days, first as curator, later as a Fellow and one of the Secretaries. In his astonishingly perceptive book, "A Discourse of Earthquakes", Hooke prepared an excellent illustration showing the morphology of ammonoids, including their suture lines. This was published posthumously, in 1705, but is said to have been written between 1686 and 1689. In the text he concluded that they were "some kind of Nautili".[31] He was way ahead of Martin Lister (1638-1711), a contemporary Fellow of the Society, who in 1671 maintained that fossils were "never any part of an Animal".[32] This was strange, because Lister was a pretty good geologist who recognized that formations with distinctive fossils could be mapped. He was another all-rounder, eventually becoming house physician to Queen Anne. Hooke's ammonoids were from English rocks, of Jurassic and Cretaceous age. There is no evidence that he handled any from the Triassic.

The literature concerning Triassic ammonoids starts with the works of two Swiss naturalists, Johann Jakob Scheuchzer (1672-1733) and Johann Jakob Baier (1677-1735). Besides writing books dealing with ammonoids they were also medical professors;

J.J. Scheuchzer

Johann Jakob Scheuchzer, F.R.S. (1672-1733) of Zurich. He was the first to describe and illustrate a Triassic ammonoid. The portrait is in his own book, "Itinera Alpina Tria", 1708.

Scheuchzer at Zurich, Baier at Altdorf. Baier's book was "Oryctographia Norica", published in 1708; Scheuchzer's was "Meteorologia et Oryctographia Helvetica", which appeared in 1718.[33] Oryctology is an archaic term for the science of fossils – paleontology. Scheuchzer described and illustrated a specimen of *Cornu Ammonis* which came to be regarded as the earliest illustration of the Triassic ammonoid now known as *Ceratites nodosus*. Of Baier's specimens, none are now believed to have been Triassic, but one, as we will soon see, nevertheless comes into the story. Scheuchzer in his early days thought that fossils were inorganic but by 1718 he regarded them as the remains of animals. Scheuchzer and Baier were not only contemporaries but also friends. Both were diluvialists, that it to say they believed that the fossils were in strata deposited by the waters of the biblical flood. Scheuchzer was a particularly ardent diluvialist, for in another book, published in 1726, he illustrated a fossil skeleton which he named *Homo diluvii testis*, which he thought was "the bony skeleton of one of those infamous men whose sins brought upon the world the dire misfortune of the deluge".[34] The French naturalist, Léopold Chrétien Frédéric Dagobert Georges Cuvier (1769-1832) would later show that Scheuchzer's fossil was a salamander, which was eventually named *Andrias scheuchzeri*.

Names like *Cornu Ammonis* and *Ammonis Cornu*, in literature published before January 1, 1758, are disregarded in scientific nomenclature as pre-Linnaean. This date is roughly that of the publication of the 10th Edition of "Systema Naturae" by the Swedish Naturalist Carolus Linnaeus (1707-1778). He invented binominal nomenclature for animals and plants, both living and fossil. In this system the minimal designation for an organism comprised two names: a genus name, followed by what is known as the trivial or specific name.

The works of the French naturalist Jean Guillaume Bruguière (1750-1799) initiate the system of nomenclature, or taxonomy, for ammonoids. In 1789 Bruguière formally recognized the genus *Ammonites*.[35] This was based on a Jurassic species. The scope of a genus is a concept. It may be interpreted to anybody's taste, as regards the individual species included within it. But it also has an objective foundation, namely the type species on which it is based. In the court of last resort, the genus is based on the type specimen (holotype) of this species. Without a type species the scope of a genus can run wild. In the old days genera were usually introduced without the designation of a type species. Later workers are entitled to straighten things out by making a designation, but with *Ammonites* there were all sorts of difficulties and now, sad to say, the name has been abandoned. But for much of the 19th century *Ammonites* was recognized as a genus, including species ranging in age from Triassic to Cretaceous.

To found a genus the scientist is required to make a diagnosis, give some comparisons, and indicate a type species. There is no rule requiring that he must have seen the type species or that this species show what are considered to be the generic characters. This is regrettable, because great and frustrating difficulties in nomenclature arise when genera are found to be based on very imperfect specimens. The most important thing is that the specimen should be a good one, well described and illustrated. Then everybody can interpret the genus. All too often in the past, and very frequently today, these conditions have not been met.

Bruguière not only named the genus *Ammonites*, but he is generally credited to have been the first to name a Triassic species – *Ammonites nodosa*, which dates from 1792. Bruguière did not provide illustrations. Instead he cited two examples. The first was one illustrated by Baier in 1708; the second, one illustrated by Scheuchzer in 1718.[36] Scheuchzer's picture had been copied in another book, "Traité des Pétrifications" published in 1742 by the versatile French scientist Louis Bourguet (1678-1742). Bruguière was referring to Bourguet's illustration when he named *Ammonites nodosa*. Because Baier's publication appeared before Scheuchzer's, and because Bruguière listed Baier's fossil first, one could

AMMONOIDS, 1708

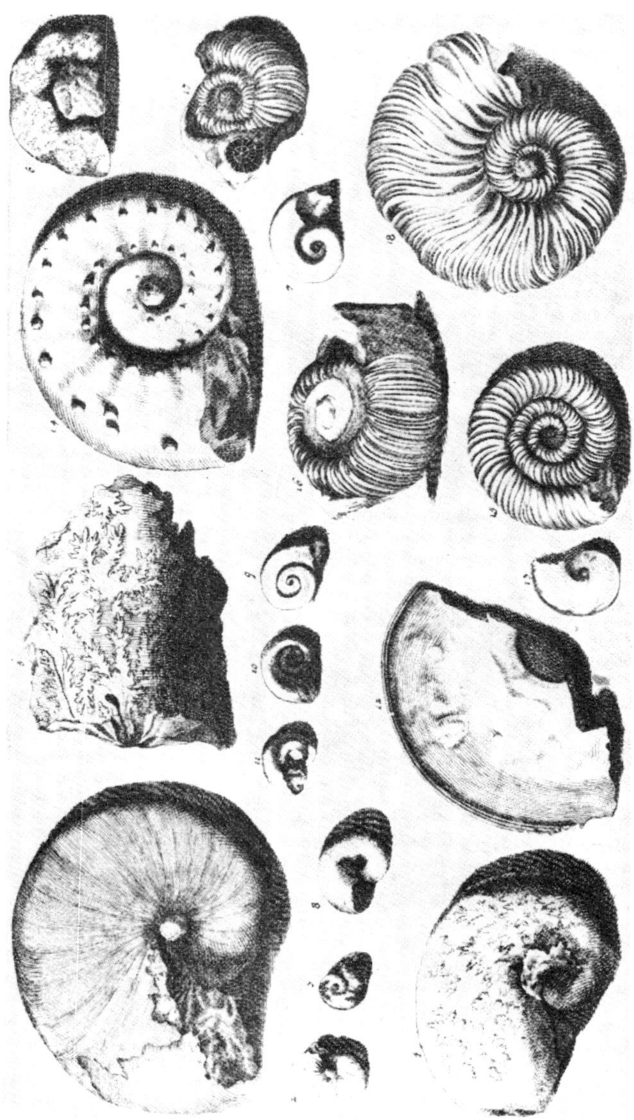

Plate of ammonoids in "Oryctographia Norica" by J.J. Baier, 1708. The original plate is about twice this size. No. 14 was described as *Ammonis cornuum verrucosam* – Ammon's horn with warts. In 1792 J.G. Bruguière formally named a species as *Ammonites nodosa*. This name is now used for a species of Triassic ammonoids – *Ceratites nodosus* (Bruguière). Bruguière gave two examples for his species: Scheuchzer's No. 25 (opposite), and Baier's specimen with warts. He thought that the two were varieties of one and the same species. In 1858 F.A. Quenstedt expressed the generally accepted opinion that Baier's fossil was a different species, of Upper Jurassic age.

AMMONOIDS, 1718

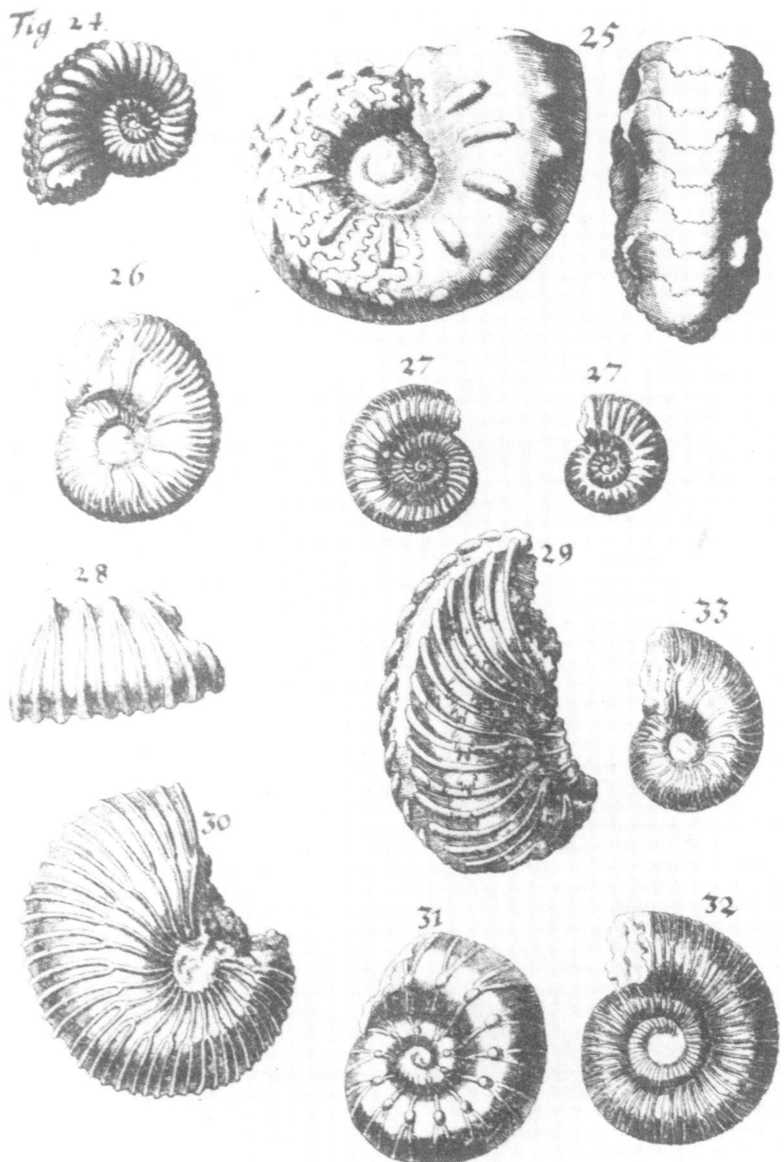

Plate of ammonoids in "Meteorologia et Oryctographia Helvetica" by J.J. Scheuchzer, 1718. No. 25 is generally regarded as the first illustration of the Triassic ammonoid now known as *Ceratites nodosus*. In 1901 Emil Phillipi described this picture as "wenig gelungen" – somewhat funny ! He suggested that the artist had put wrinkles on the suture line saddles which were not really there.

argue that *Ammonites nodosa* should be interpreted from Baier's fossil, not Scheuchzer's. Both pictures are not very good, but good enough to show that the fossils were not the same according to modern standards. Rightly or wrongly Scheuchzer's picture, not Baier's, came to be regarded as definitive, but even so it is impossible to say what the species really looks like from the 18th century literature. If one chose to interpret the species from Baier's illustration the consequences would be disruptive, to say the least, because Baier's specimen was thought to be a Jurassic ammonite by F.A.Quenstedt ! *Ammonites nodosa* is nowadays interpreted from a specimen that was unknown to Baier, Scheuchzer and Bruguière. Although this may be illegal any attempt to make a change would cause nothing but confusion.

The next person involved in the description and illustration of Triassic ammonoids was Franz Xavier Freiherr von Wulfen (1728-1805). Wulfen was a professor of Physics, Mathematics and Philosophy at Klagenfurt. He wrote a book, illustrated with hand-coloured plates, with descriptions of ammonoids from Bleiberg in Austria.[37] They are now known to be Carnian. Wulfen seems to have been unaware that Bruguière had proposed the genus *Ammonites*. He assigned his specimens to the genus *Nautilus* which had been recognized by Linnaeus in the 1758 "Systema Naturae". Two species, now known as *Carnites floridus* and *Joannites cymbiformis*, are attributed to Wulfen on the assumption that he was using the Linnaean binominal system when he published his book in 1793. Wulfen's style of presentation does not consistently conform to this system and the use of his names could be questioned. This was the conclusion of Charles Davies Sherborn (1861-1942), the scholarly and painstaking paleontologist who worked at the British Museum and made a monumental Index of all the zoological names proposed before 1800. But as with *Ammonites nodosa* this seems to be another nomenclatural sleeping dog best left undisturbed.

An attractive little book on ammonoids was published in 1818 by Johann Christoph Matthias Reinecke (1769- 1818). He, like Wulfen, used the genus *Nautilus* instead of *Ammonites*. He described the species *Nautilus undatus* and provided a beautiful hand-coloured illustration. He also gave the locality where it was found. Although the actual specimen has not been traced there is no doubt that it was a Muschelkalk *Ceratites*. Some workers have considered that Reinecke's species is synonymous with *Ceratites nodosus*; others regard the two as distinct species. As Sherborn remarked in the introduction to his Index, it "depends on the idiosyncrasy of the systematist". Reinecke's book attained a much higher and more useful standard than those of his predecessors. He became the "Gymnasial-direktor" in Coburg and evidently gave a lot of thought to these fossils because he has been proclaimed one of the earliest champions of evolutionary theory.[38]

In 1825 Guilielmo (Wilhelm) de Haan (1801-1853), Curator of the Leyden Museum, created the genera *Goniatites* and *Ceratites*.[39] Soon after he was followed by von Buch, who, between 1829 and 1848, organized the ammonoids into three groups: Goniatiten, Ceratiten and Ammoniten.[40] Ammonoid shells are coiled cones. The cone is divided into chambers by partitions known as septa. The animal lived in the last chamber. The chambers in the early-formed part of the shell were originally occupied by the animal, but as it grew and moved on became filled with gas, providing buoyancy. Von Buch made use, for classification, of the nature of the contact between the septum and the inner side of the shell wall. In all ammonoids this is folded, with the resulting folds being called saddles if convex forwards, lobes if concave. In some, for example *Pinacoceras*, the resulting pattern (the suture line) is extraordinarily elaborate. Von Buch seems to have been the first to realize that the Mesozoic ammonoids had more complicated suture lines than those of the Paleozoic. In taxonomy, however, he was a conservative. "Ceratiten" he recognized, but the genus *Ceratites* he did not.

So, by 1825 there were three named genera available for Triassic ammonoids: *Goniatites* if both lobes and saddles were unfrilled; *Ceratites* with denticulation of lobes alone; and *Ammonites* with frilling or toothing of both lobes and saddles. Well into the 19th century these three were to be used for all species of Triassic ammonoids. *Goniatites* is nowadays restricted to Paleozoic species; *Ceratites* is still an accepted Triassic genus; *Ammonites*, as we have seen, has been ditched by the Jurassic specialists.

From 1860 many new generic names began to appear. The creation of new genera and the naming of dozens of species continues to this day. The elaborate resulting nomenclature (taxonomy) is admittedly cumbersome. Unfortunately this aspect of ammonitology turns many away from an exciting and fruitful field of research. The elaborate taxonomy is necessary, however. I will attempt an apologia. Ammonoids evolved rapidly and produced an astonishing variety of forms. Many genus names are necessary to label and characterize this great morphological variety. The curious feature of ammonoid evolution is that it is difficult to see what was achieved. From the shells alone we have no reason to suggest that the animal inside the shell of a Devonian *Clymenia* was appreciably different from the inhabitant of a Permian *Xenodiscus*, a Triassic *Ophiceras*, Jurassic *Psiloceras* or Cretaceous *Lytoceras*. Among the cephalopods, dramatic evolutionary achievements are recorded in the history of the Coleoids, the group that includes the living squids. Ulrich Lehmann has put it well: coleoids became the equivalent of jet aircraft compared with ammonoids, which, with their buoyant external shells, were balloons.[41] They were balloons in the Devonian; and remained balloons until their final extinction at the end of the Cretaceous. They must have been quite successful balloons, judging from their abundance and longevity, but it would seem that they eventually gave way to other more agile and versatile marine creatures.

An interesting aspect of ammonoid evolution is its repetitive nature. This is best illustrated by the heteromorphs – ammonoids that depart from the normal planispiral form and instead are coiled like snails, or are not coiled at all. They evolved in the Triassic. They are also found in the Cretaceous. There seems to be no doubt whatsoever that these heteromorphs of greatly different age evolved quite independently. To use some ammonoid jargon, they are "heterochronous homeomorphs". Know ye all men ! Heteromorphs are heterochronous homeomorphs ! This sort of evolution evidently also took place at more prosaic levels and herein stem many of the difficulties with ammonoid taxonomy. Two ammonoids look more or less alike; does this indicate a close biological relationship or not ? The taxonomic systems used today are designed to provide different names for ammonoids which look roughly or almost exactly similar, but which evidently arrived at the similar form quite independently.

Distinguishing heterochronous homeomorphs admittedly may be a tricky business. Arguments have often developed. The most notable was between Gustav Steinmann and Carl Diener early in the 20th century.[42] Steinmann (1856-1929), for much of his career a professor at Bonn, was a geologist and paleontologist of astonishingly wide interests.[43] He knew a lot about ammonoids, but was also very knowledgable about ultrabasic rocks, radiolarites and pillow lavas. He noted the frequent association of the last three in Alpine-type mountain belts and was thus responsible for the ophiolite concept. This was sanctified as the "Steinmann Trinity" by Sir Edward Bailey (1881-1965).[44] Steinmann's observations now play an important role in the interpretation of global tectonics, his Trinity being generally interpreted as the tectonically tortured remains of oceans that have now been squeezed out of existence. When it came to ammonoids, Steinmann was the advocate of "Rassenpersistenz" – persisting races. *Pinacoceras aspidoides* (Middle Triassic) looks like *Oppelia aspidoides* (Middle Jurassic). To Steinmann these two were representatives of a long-lived race. Diener disagreed. If Steinmann is right, why are there none in the Upper

CEPHALOPODS, 1818

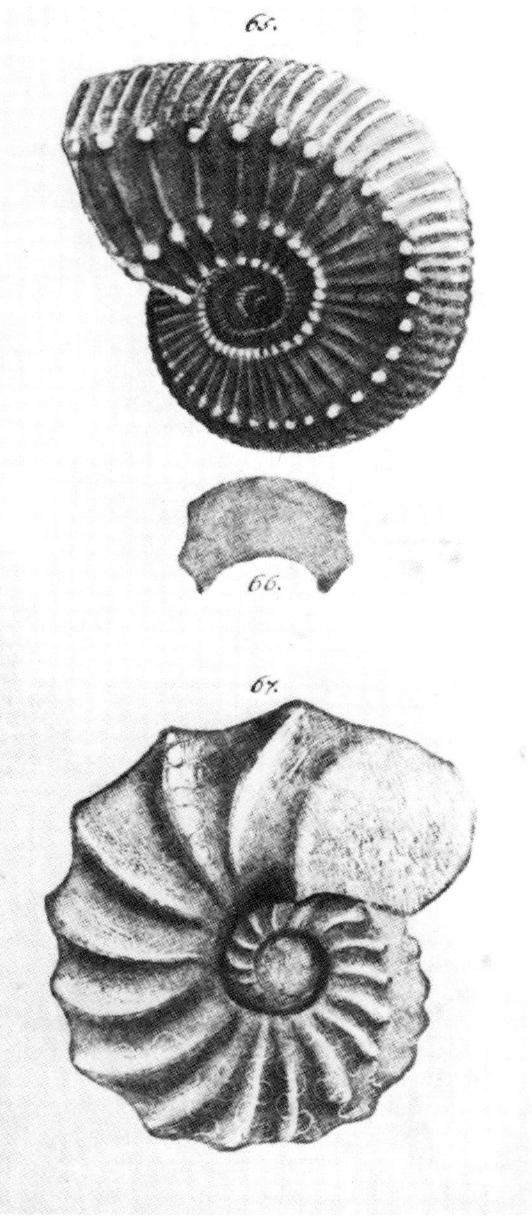

Plate VIII from the book on fossil cephalopods by J.C.M. Reinecke, 1818. No. 67 was named *Nautilus undatus*. It is now recognized to be a species of the genus *Ceratites*. The ceratitic suture line is clearly shown. It came from the Muschelkalk (Middle Triassic) of Langen Berg, north of Coburg, Germany. On the original plate the specimen is painted grey in water colour. This is the usual colour of Muschelkalk fossils. Nos. 65 and 66 are views of a Jurassic ammonite.

Triassic and Lower Jurassic ? Virtually everybody agreed that Diener won the argument and Steinmann's lucid advocacy for his case is not accepted today. Instead, Steinmann's argument is taken to illustrate the monotonous and repetitive nature of ammonoid evolution.

Zoologists dealing with living animals have many significant data – breeding habits, accurate knowledge of geographic range and habitat, as well as morphology, which contribute to their definition of species. With ammonoids we generally only have morphology, and often not much of that. We have one great advantage over the zoologist, however, namely precise and accurate data on the sequence of the faunas over periods of millions of years. So the problems are different. It ill behoves zoologists to tell ammonitologists how to handle their taxonomy.

Perhaps many ammonoid genera were species; escalating the taxonomy so that the genus is the main taxonomic category may be wrong. But it must be remembered that the definition of an ammonoid species, on the supposition that the rigid criteria and constraints employed by zoologists can be applied, is impossible. Incomplete ammonoids – ones without body chambers or in other ways a poor representation of what the complete shell of the living animal may have looked like – are very common; the rule rather than the exception. But these incomplete specimens are nevertheless amenable to profitable scientific study. Incomplete ammonoids with significant similarities may be found to characterize a short interval of time in say Taymyr, Spitsbergen and Ellesmere Island. But they may not be exactly the same. Minor differences of uncertain significance are best recognized by giving the specimens different trivial names, that is to say, calling them different species. Similarities judged significant find formal recognition by placing them in the same genus. The many different names given to ammonoids are thus useful and necessary, particularly as it is found that many genera, defined on restrictive morphological criteria, only lasted for one or two million years.

In the 1950s an interesting dialogue began to develop between William Jocelyn Arkell (1904-1958) and the eminent living zoologist Ernst Mayr, of the Museum of Comparative Zoology (MCZ) at Harvard.[45] Arkell was a Jurassic ammonitologist with immense practical experience. Mayr is the doyen of systematic zoologists. It is a pity that the dialogue was cut short by Arkell's untimely death. Arkell found it best, in order to express his findings on the morphological variety shown by his Jurassic ammonoids, to adopt what is described as the typological system. With this, ammonoids with distinct morphological characters are given specific names different from those of their fellows, even if, in some cases, they seem to be linked together by morphologically intermediate forms. Mayr objected that Arkell failed to adduce proof "that these different morphological types are really different and reproductively isolated populations". Arkell's species, and even genera, he held, might be ecophenotypes – variants – of one species. A slight paraphrase of Arkell's reply: "Of course there is no proof; what matters is the existence of a wide range of forms only expressable by the use of specific names within the framework of a chosen taxonomic scale." As an ammonoid taxonomist Arkell was surely right. There is no possibility of adducing proof of interbreeding. The definition of ammonoid species may be judged arbitrary and inconsistent by zoologists. I would not disagree. But it does not matter. A taxonomic scale in which the genus is pre-eminent serves the ammonitologist best. For a genus a satisfactory diagnosis is attainable. For a species it is seldom so. Some ammonoid species may be ecophenotypes; some may be subspecies; others may have been true species. Proof cannot be adduced. Rules for living animals are not for ammonoids.

Bernhard Kummel (1919-1980), an active worker on Triassic ammonoids for many years, was a colleague of Mayr's at the MCZ. Kummel was much concerned with the species problem. He chose to interpret many collections on the assumption that Mayr's ecophenotype ideas were right. Sometimes Kummel placed a dozen or more previously

CANADIAN TRIASSIC AMMONOIDS

Seventy-four different Triassic ammonoids from Canada, about one third natural size. They range in age from earliest to latest Trias, older ones at the bottom, younger at the top. Bottom left is *Otoceras. Trachyceras* is half way up on the right. The biggest is *Distichites*. Seven o'clock to *Distichites* is *Nathorstites*, to the left of which is *Daxatina*, originally known as *Dawsonites*. The small ones at the top are heteromorphs, a loosely coiled *Choristoceras, Cochloceras* coiled like a snail, and the straight *Rhabdoceras*. A complete key is on page 153

named species in synonymy with the first named.[46] This meant that ammonoids that looked very different ended up with the same name. Perhaps Mayr and Kummel are right, but they, like Arkell, have not adduced proof. In the past I have flirted with the ecophenotype, as opposed to typological approach myself, but I now think that when the name of an ammonoid ceases to provide a reasonably restrictive indication of what it looks like, it is not useful, for biochronology or anything else. But I do not deny the difficulties; it is seldom that two ammonoids are exactly alike. It is probably not a good idea that every specimen should be a different species ! This is what virtually came about when Mojsisovics and Waagen wrote their monographs in the 1890s. But it is worth remembering that these books remain immensely valuable for identifying ammonoids.

The status of an ammonoid species is not worth a quarrel. F.H.McLearn, an important figure in the later part of this story, told me that this was the advice given to him by Professor Schuchert of Yale. Quarrel about a genus; this may be worth while, but not about a species ! I am not inclined to quarrel much about genera either. I would only quarrel with those who create genera without knowing exactly what the type species looks like.

All this should answer claims that the large number of names – species and genera – that encumber the ammonoid literature are unnecessary. As John Callomon has said, the burdensome nomenclature is necessary to deal with a truly complicated subject.[47]

Returning now to the mid 1800s. Some of the new names that appeared were necessary to characterize truly new discoveries. Hauer, who was pretty conservative in his use of *Goniatites, Ceratites* and *Ammonites*, produced two excitingly new genera by being the first to describe Triassic heteromorphs – *Rhabdoceras* (straight) and *Cochloceras* (snail-like) in 1860. Suess, in 1865, wrote a very perceptive paper entitled "Ueber Ammoniten". In this he recognized some fundamental evolutionary lines among the forms until then placed within the genus *Ammonites*. The genera *Arcestes*, *Phylloceras* and *Lytoceras* were named and *Ammonites* was given a more restricted scope. Suess' divisions have fared pretty well, to the extent that his genera are now elevated to about the level of Suborder ! *Lytoceras* and its allies are younger than Triassic; *Phylloceras* is now restricted to Jurassic forms. But it is still recognized to have close relatives in the Triassic fauna. This was Suess' masterly stroke. He saw *Phylloceras* as the only link between the Triassic and Jurassic; the survivor that protected all the ammonoids from extinction. Suess' ideas still arouse controversy. I now agree with Jean Guex that he was wholly right.[48]

An important spiny ammonoid – *Trachyceras* – was split from *Ammonites* by the Prague Professor, Gustav Laube (1839-1923),[49] in 1869. In 1873 Mojsisovics took the gigantic disc-like ammonites described by Hauer as *Ammonites metternichi* into the new genus *Pinacoceras*. *Pinacoceras* must have been a sight to see. The chambered part attained a diameter of at least 700 mm but was only about 100 mm wide. In life it must have been about one metre high, very thin, smooth, with an edge much like a knife. The suture was frilled to make an astonishingly beautiful pattern, but this would not have been visible in life. When Mojsisovics first defined the genus, he started with the words "animal unknown", expressed, perhaps, with a tinge of regret. Nobody seems to have objected to Mojsisovics' observation, but when F.N.Johnston made the same remark about an ammonoid from Nevada, in 1941, it evoked a distinctly patronising, indeed sarcastic, comment from Spath.[50] At least Spath could have picked on Mojsisovics as well !

From 1875 the decimation of both *Ceratites* and *Ammonites* gathered momentum, mostly through the endeavours of Mojsisovics, Waagen and Alpheus Hyatt. Soon *Ammonites* no longer belonged in the Triassic fauna; *Ceratites* survived and does to this day. By 1900, when Hyatt prepared the English-language edition of Zittel's famous text book, there were 164 genera. The latest census lists 445,[51] and there are still not enough.

PIONEERS

Leopold von Buch
(1774 – 1853)

Friedrich August von
Alberti (1795 – 1878)

Eduard Suess
(1831 – 1914)

Franz von Hauer
(1822 – 1899)

IV

Trias Beginnings – Germanic and Alpine – Events to 1850

The typical or Germanic Trias, comprising the Bunter, Muschelkalk and Keuper formations is best developed in the part of Germany bounded by the Ardennes, Vosges, Black Forest and Bohemian Mountains. Bunter (literally "coloured") and Muschelkalk (clam-limestone) had been recognized in the 18th Century by Lehmann and Füchsel (Chapter I). Keuper was named later, in 1822, at the suggestion of von Buch. It was apparently the name used for the terrain by the inhabitants of Coburg. Formal proposal of the name Trias for these three was made in 1834 by Friedrich August von Alberti (1795-1878).[52] The Muschelkalk has a varied marine fauna; the other two are mostly red non-marine strata. At the top of the Keuper, as interpreted by Alberti, was his Täbingen Sandstone with a marine fauna characterized by *Avicula contorta*.[53] These beds are now known as the "Rhät" in Germany. In England the Muschelkalk is missing. There most of the package became known as the New Red Sandstone followed by what used to be called the "Rhaetic". The British "Rhaetic" has now been named the Penarth Group. The Muschelkalk fauna (dominated by the ammonite genus *Ceratites*) has a provincial character, probably reflecting abnormal, hypersaline, conditions of deposition. For these reasons – absence of marine faunas in the lower part; an abnormal fauna in the middle; and a marine fauna without ammonoids at the top, – the typical Germanic Trias does not represent an ideal standard against which correlatives can be recognized worldwide.

At about the same time as the typical Germanic Trias was being studied and named, geologists were also wrestling with the much more difficult geology of the Alps. Fossils were discovered early on, but the sequential relations of the rocks in which they were found were not clear, and in some places are still not clear. Nearly all the fossils were new and different from anything known in the more northerly parts of Europe, where the Trias was obviously sandwiched between the Permian and the Jurassic.

Many well known geologists and paleontologists worked in the Alps in this period. They included the English geologists Roderick Impey Murchison (1792-1871) and Adam Sedgwick (1785-1873); also von Buch. These pioneers recognized an extensive "Alpine Limestone". Sedgwick and Murchison toured the Alps together in 1829. This was in the days when they were friends and collaborators, long before the dispute about the limits of the Cambrian and Silurian systems which led to their permanent estrangement. In 1835 they published a geological map of the eastern Alps showing the extent of the Alpine Limestone.[54] Sedgwick and Murchison thought that the limestone, most of which is now known to be Triassic, was wholly Jurassic. This illustrates how difficult it was to correlate between the Alpine and Extra-Alpine areas in those days.

Fossils from the Alpine Limestone were described early on. Some clams that would assume great importance were named *Pectinites salinarius* by Ernst Friedrich Baron von

GERMANIC AND ALPINE TRIAS

SALZKAMMERGUT
Lake Balaton
Rhine R.
Rhone R.
Danube R.
HAN BULOG
DOBROGEA
BLACK SEA
KOKAELI PENINSULA
DOLOMITES
PALERMO
EPIDAUROS
CHIOS
MEDITERRANEAN SEA

Patterned area has typical Bunter, Muschelkalk and Keuper.
Named are some important localities for the Alpine Trias.

500 miles
500 km

Schlotheim (1765-1832) in 1820. Schlotheim described these fossils but gave no illustrations. One wonders how the early 19th century paleontologists, like Schlotheim, expected to communicate with their scientific colleagues when they described fossils without giving any illustrations. It was certainly common practice, but in those days the scientific community was small and everybody probably knew everybody who was anybody. So when Schlotheim described *Pectinites salinarius* presumably everybody who mattered was supposed to know what he meant. Perhaps he expected that they all had an example in their cabinet ! These fossils became better known to scientists at large in 1830 when Heinrich Georg Bronn (1800-1862),[55] of Heidelberg, decided that Schlotheim's name was being given to two different clams. He named them *Halobia* and *Monotis*, and provided illustrations. Now we had *Monotis salinaria* (Schlotheim) and *Halobia salinarum* Bronn described and illustrated. These two, as the 19th century advanced, assumed an immensely important role for recognizing the Trias worldwide. But in 1830 their age was unknown. This was still true when they were figured in the famous index fossil book of the time, Petrefactae Germaniae, published by Georg August Goldfuss (1782-1848)[56] between 1834 and 1840. They are not found in the Germanic Trias, where the beds in which they might be expected are the non-marine Keuper. This illustrates the nature of the dilemma: fossils suspected to be Triassic were found in the Alps where the stratigraphy was far from clear. The Germanic sequence provided a clear sequence, but few important fossils.

The etymology of these fossil names poses a curious question. "Salinaria" implies salt and clearly alludes to the occurrence of the fossils near to the famous salt deposits of Salzkammergut and Hallein, which were known to Iron Age Man, at least 5 centuries B.C. But what did Bronn have in mind when he named *Halobia* ? Was he alluding to the fact that the fossil was found near a salt deposit ? Or was he, the explanation I would prefer, naming it as an extinct creature that had lived in the salty sea ? *Halobia* presumably can be construed as either living creature found near the salt or living creature of the sea. Perhaps it was an intentional double entendre. Bronn doesn't say. In passing it may be mentioned that there is a vessel, "Halobia II", registered under the Canadian Shipping Act (ON 346256, ex "Karim IV"). This vessel was named on the assumption that the second etymological construction was in Bronn's mind.

The Alpine Trias is now known to be very extensive, particularly in Austria and Italy, with some important localities in southern Germany (Bavaria) and Switzerland. There are two areas of particular historical significance which will be referred to repeatedly in the pages that follow. One is Salzkammergut, the surroundings of Hallstatt Lake, in the northern Austrian Alps. The other is to the south, in the area now known as the Dolomites wherein lies the important locality of St. Cassian. Nowadays it is San Cassiano, an Italian village. In the last century the Dolomites were in South Tirol, a part of the Hapsburg Empire.

In preparation for the problems and controversies that are brought out below, it may be said that Salzkammergut is an area where the stratigraphy and geological structure is very complex and difficult to interpret. But for Triassic ammonoids it is about the richest source anywhere in the world. Most of them come from the Hallstatt Limestone and the Zlambach Marl; a few from the Pötschen Limestone. Salzkammergut ammonoids are objects of extraordinary beauty. Those from the Hallstatt are red or grey; those from the Zlambach, grey. Some of the red ones are coated with a film of almost black manganiferous mineral. Ammonoids with red insides and black exteriors are natural objets d'art. Some of the grey ones preserve the suture lines clearly and make attractive objects when polished. Early on, Salzkammergut ammonoids were sought by professional collectors and sold as attractive curios. This sort of activity continues to this day, but the specimens obtained are not always for sale. In some ways science has been advanced by the work of these collectors; in others,

FOSSIL CLAMS AND AMMONOIDS OF THE ALPINE TRIAS

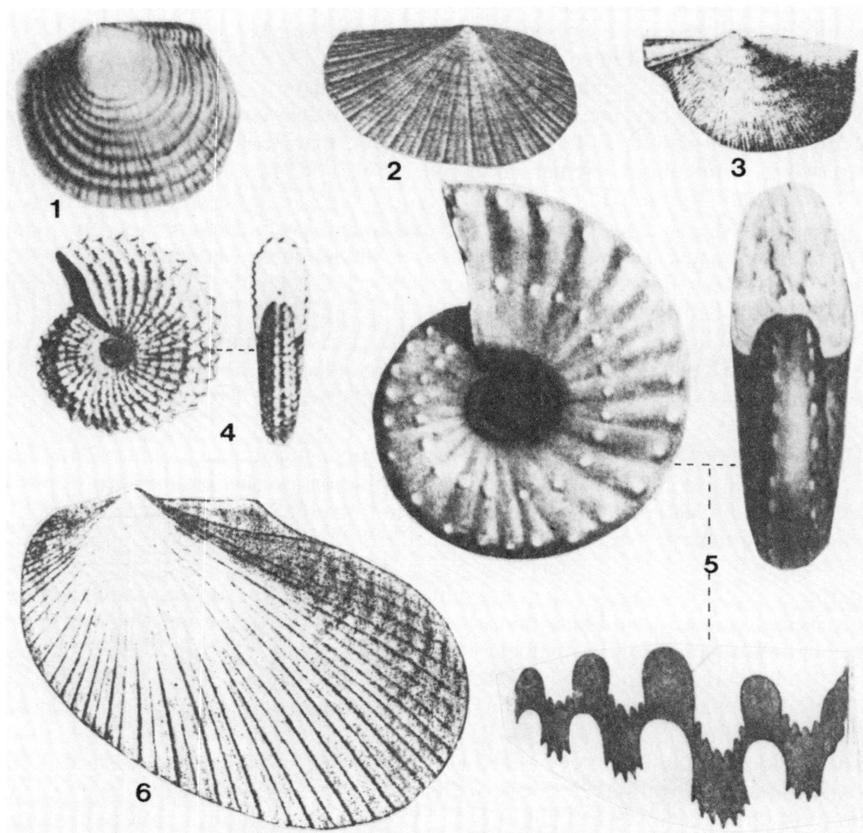

1. *Posidonomya clarae* Buch; 2. *Halobia lommeli* Wissmann; 3. *Halobia salinarum* Bronn; 4. *Ammonites (Ceratites) aon* Münster; 5. *Ceratites binodosus* Hauer; 6. *Monotis salinaria* (Schiotheim). 1 and 5 were illustrated by Hauer in 1851; 2 by Wissmann in 1841; 3 and 6 by Bronn in 1830; 4 by Münster in 1834.

as we shall see, grievously hindered (Chapter XVI). In the last century the professional collectors sold specimens to the Vienna scientists. Although these specimens have been preserved for posterity, sometimes their full scientific significance cannot be assessed because the collectors did not give exact details as to where the fossils were found. "Would not" perhaps is more the truth; a professional collector might be expected to guard jealously the exact location of his mine. The Austrian statesman Metternich had a good collection which was described early on by Franz Ritter von Hauer, about whom more later (Chapter V).

Nearly all the Salzkammergut localities are within a radius of about 15 km of Hallstatt town and had been discovered by the middle of the 19th Century. Two important localities, Sommeraukogel and Steinbergkogel are near the salt mines immediately above the town. Millibrunnkogel and several other localities are west of Sandling Mountain, on the north side of the Leisling Valley, some 10 km to the north. The Pötschen Limestone locality is nearby. "Kogel" is the Austrian word for a round-topped mountain or hill. On festival occasions fires were lit on some of these hills. Such a hill is known as a "Feuerkogel". Some Austrian maps show lots of Feuerkogels. One particular Feuerkogel, near Röthelstein Mountain, about 15 km east of Hallstatt, is perhaps the most famous Triassic ammonoid locality in the world. Although these localities were known early on, until the 1920s they provided virtually no reliable data on the succession of ammonoid beds. Really reliable data did not come until the late 1960s, from the work of Leopold Krystyn and his colleagues (Chapter XVI). Salzkammergut then, although a boon to the collector of ammonoids was a disaster area for the elucidation of stratigraphy in the early days. The problems of Salzkammergut stratigraphy, particularly whether the Zlambach Marl was above, below, or equivalent to the Hallstatt Limestone, were the root of the polemics that engulfed the whole Austrian geological community in 1898, a matter so involved that it will have a Chapter (VIII) of its own.

The Dolomites, on the other hand, are a paradise for not only the paleontologist, but also for the stratigrapher and students of reefs and volcanic rocks. Ammonoids occur in the St. Cassian beds and at several lower levels. Exposures are excellent and many stratigraphic relationships are clearly demonstrable. By the middle of the 19th century the sequence in the Dolomites was pretty well known.

In the early days there were two big problems with the Trias. First there was the problem of correlation within the Alps, e.g. between Salzkammergut and the Dolomites. Then there was the problem of correlating between the Alps and the typical Germanic development. Not much progress was made until the description of the St. Cassian fauna, in 1834, by George Graf Münster (1776-1844). This was partly from collections that had been made by von Buch. Münster lived in Bayreuth where he was an official of the Bavarian government. He accumulated a large collection of fossils which eventually became the nucleus of the Bavarian State Collection in Munich, where they can be seen today. Nearly all the fossils he described were new to science. Münster nevertheless noted that a few in the rich and varied St. Cassian collection suggested affinity with the Muschelkalk fauna. This was the first step in correlating between the Alpine and Extra-Alpine areas. Among the many fossils that Münster described, the most important could not be said to indicate affinity with the Muschelkalk fauna. This was *Ammonites aon*, a species soon to play a very important role in making correlations, both with Salzkammergut and the Himalayas. A further important contribution to the faunas of the St. Cassian area was made by Heinrich Ludolf Wissmann,[57] in a paper published jointly with Münster in 1841. Wissmann was born in Göttingen in 1815 but nobody seems to know when and where he died. He described *Halobia lommeli*. *Halobia*, as noted above, was one of the fossils described from the undated limestone of the northern Alps by Bronn. Wissmann recognized that his fossil and

TRIAS OF DOLOMITES

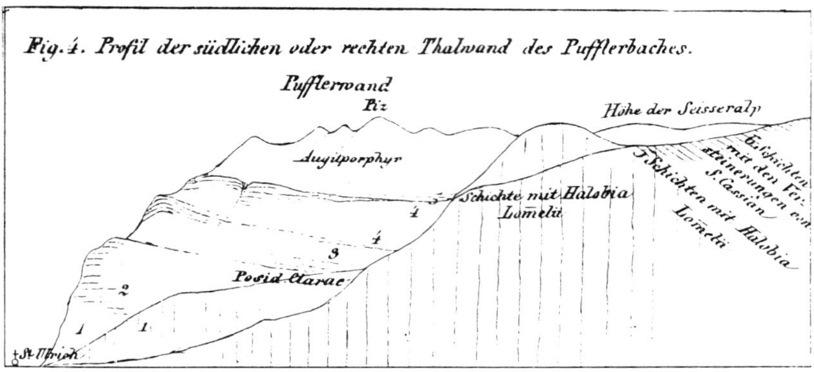

Above: Interpretation of stratigraphy of south side of Val Gardena, given by H.F. Emmrich in 1844. Succession of Lower, Middle and Upper Triassic faunas is shown. Below: contemporary students of the Carnian St. Cassiano beds (Rinaldo Zardini, Jobst Wendt, Max Urlichs) standing on the formation at Alpe di Specie. Mountains behind are of Dolomia Principale (Norian).

Bronn's were not exactly the same; his is now known as *Daonella*. But the similarity was judged significant by Wissmann, and was very important because it provided the first clue towards correlating the limestone in the northern Alps with the strata around St. Cassian.

A little later, August Wilhelm Klipstein (1801-1894)[58] wrote about the St. Cassian fauna. Klipstein's work was regressive. He mistakenly thought that the St. Cassian beds were underlain by strata containing a Jurassic ammonite.[59] He thus came up with the idea that the St. Cassian beds were younger than Trias, which was wrong.

Before leaving this place, at this time, one must mention the important contribution of Hermann Friedrich Emmrich (1815-1879).[60] In 1844 he gave a remarkably clear and accurate description of the sequence in the Dolomites, showing that there were at least three faunas in sequence; the lowest with *Posidonomya clarae*, the next with *Halobia lomelii* (sic) and a third with the St. Cassian fauna (*Ammonites aon* etc.).[61] Hopefully the contemporary savants knew what he meant by *Posidomomya clarae* because no illustration was published for another six years. This clam, now in the genus *Claraia*, later turned up worldwide, marking a fauna very close to the base of the Triassic. Emmrich thus laid the foundation for recognizing what would eventually be the Lower, Middle and Upper Alpine Trias. But nobody wins them all ! We will see in the next Chapter that Emmrich was not so successful in deciphering the geology of the northern Alps.

Von Buch, so knowledgeable about all aspects of geology and paleontology, wrote an interesting paper – Ueber Ceratiten" – in 1848. This may be taken to mark the end of the pioneer age. It was a landmark because he concluded that ceratitic ammonoids characterized an interval of time which he could recognize in widely scattered parts of the world. More will be said about this in Chapter VI. Von Buch exhibits a curious conservatism. Alberti's name Trias, although more than 10 years old, was not used; nor was the genus name *Ceratites* (dating from 1825) employed. But this in no way detracts from the fruitfullness of his ideas.

PINACOCERAS FROM THE ALPINE TRIAS

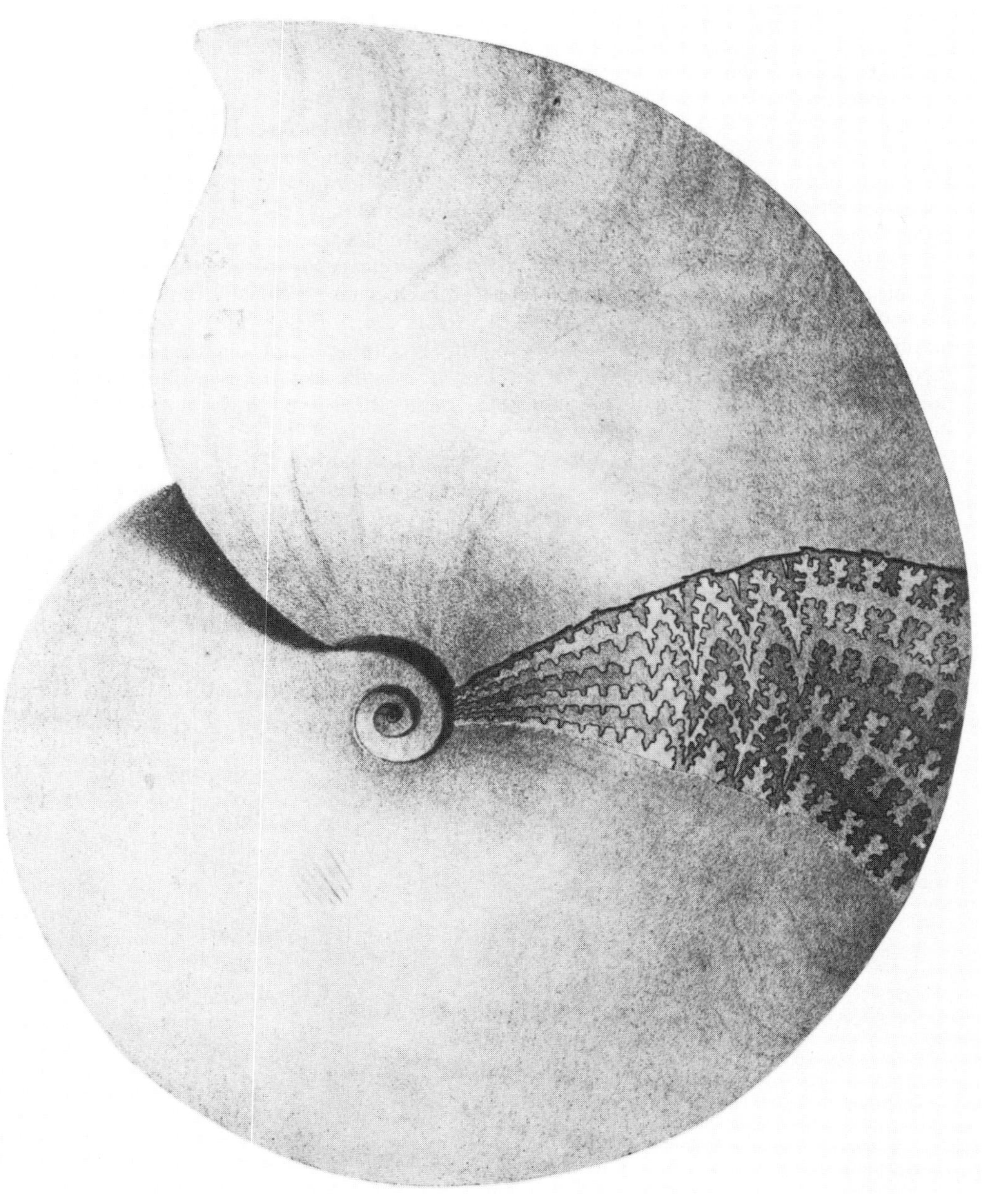

Hauer's illustration (1846) of *Ammonites metternichii* from the Sommer-aukogel, near Hallstatt in Salzkammergut, Austria. This ammonoid is now placed in the genus *Pinacoceras*. The diameter was about 210 mm, width about 27 mm. The edge was sharp.

V

Hauer and the mid 19th Century Synthesis

The central figure of this period was Franz Ritter von Hauer (1822-1899), eventually described as the doyen ("der Nestor") of Austrian geologists.[62] Between 1846 and 1866 he was very active in research on the Alpine Trias. Then, between 1867 and 1885, he became Director of the Imperial – Royal Geological Survey (Kaiserlich – Königlichen Geologischen Reichsanstalt), which had been founded in 1849. During this period, besides running the Reichsanstalt he wrote a classic book on all aspects of the geology of the Austro-Hungarian Monarchy.[63] In 1885 he was succeeded as Director by Dionys Stur (1827-1893), who had worked extensively on the Alpine Trias in the 1860s. Hauer then became Director of the Natural History Museum in Vienna. There he returned to ammonoid research by studying the important Middle Triassic faunas from Han Bulog and Haliluci in Bosnia. He was recognized by election to the Imperial Academy in Vienna; also internationally by election to many foreign academies and in 1882 by the premier award – the Wollaston Medal – of the Geological Society of London. The funds for the minting of this medal in gold had been provided by the eminent mineralogist, William Hyde Wollaston (1766-1828), who discovered the element palladium. The first recipient, in 1831, was William Smith. Between 1846 and 1860 the medal was struck, appropriately, in palladium that had been donated to the Geological Society. But when von Hauer received it they had run out of palladium and were using gold again. Nowadays they are back to palladium.

Hauer made outstanding contributions. He was the first to describe many of the ammonoids that would eventually become the index fossils of Triassic zones. He dealt with faunas from all parts of the Triassic, from the oldest to the youngest, describing material from both the northern and the southern Alps. Particularly important was his recognition in the Hallstatt Limestone of the northern Alps, of ammonoids resembling those described by Münster and Klipstein from the St. Cassian beds. This provided firm grounds for north-south correlation within the Alpine belt. An atlas incorporating all the fossils first made known by Hauer would make a pretty good guide for recognizing many of the zones recognized worldwide today. His taxonomy was conservative. Most of his species were assigned to *Ammonites*; some to *Goniatites*. But he also recognized *Ceratites*, and thus gave grounds for correlating between the Alpine area and the Muschelkalk. So it came about that the term Alpine Muschelkalk came into use. He also worked on clams. The first illustrations of the important species now known as *Claraia clarae*, the worldwide Lower Triassic index fossil, were provided by Hauer. He did not claim to be the first to know it: it was described as "*Posidonomya clarae* von Buch." This seems to be one of the fossils that was known to everybody long before it was figured; Emmrich, as we have seen, also knew about it. Hauer was also the first to describe *Choristoceras*, the youngest Triassic ammonoid of all.

Many of the fossils that Hauer described were collected by other people. His first ammonoid paper, published in 1846, was on specimens from the Hallstatt Limestone. They were in a collection belonging to the Austrian statesman, the Fürst von Metternich (1773-1859), who held so much of Europe with a tight rein between 1815 and 1848. I don't think Metternich had hammered them out of the rock himself; probably he bought them. But it is on record that Metternich was seriously interested in geology. In 1830 he had a conversation with Sir Roderick Impey Murchison, claiming that he really preferred science to politics. He said that it was only at the insistence of his father that he took up the latter. Murchison recounted this conversation to Archduke John, who said that "it was all fudge and merely intended to blind me".[64] Within two years of Metternich's ammonoids being described, much of Europe, including Austria, was convulsed in the revolutions of 1848, although none were as severe as the earlier one in France. Metternich was far from popular. He resigned, and retreated to England with an armed escort. He later returned to his castle on the Rhine but never resumed politics. I don't know what happened to his ammonoids. Apparently they were not available to Mojsisovics when he was revising the Hallstatt fauna, some 25 years later.

Rather surprisingly the *Choristoceras* described by Hauer was found by an American, Othniel Charles Marsh (1831-1899), who later became the famous vertebrate paleontologist, Yale professor, and antagonist of Edward Drinker Cope (1840-1897), who also worked on vertebrate fossils. An account of the quarrels between Marsh and Cope has provided enough to fill at least one book.[65] We will soon see that nearly a book would also be needed to describe fully the feud that developed between two of the Trias personalities, Mojsisovics and Bittner. Marsh found the fossils while on a European educational tour in the 1860s. We have already seen (Chapter III) that Hauer was also first to describe the Triassic heteromorphs – *Cochloceras* and *Rhabdoceras*. These are only a few of the things that he made known.

But Hauer was no arm-chair paleontologist. In 1850 he published an exceedingly important paper in which he discriminated Triassic and Jurassic beds within the Alpine Limestone,[66] the formation that Sedgwick and Murchison mistakenly thought was wholly Jurassic (Chapter IV). It was another erroneous interpretation, by Emmrich, that prompted Hauer to make the correction. In doing so he seems to have been the first to extend use of the term Trias into the Alps. *Monotis salinaria*, until then in limbo, was recognized as a guide to the Upper Trias. Hauer also gave new evidence for correlating some of the northern limestones with the St. Cassian strata, which supported Wissmann's interpretation (Chapter IV). In 1856 the Vienna paleontologist Moritz Hoernes (1815-1868), who was Suess' brother in law, also provided data contributing to this correlation.

These were giant leaps forward. One must remember that only five years earlier Klipstein was insisting that the St. Cassian Beds were Jurassic (Chapter IV). There is a lot of interesting reading about the different views on the Alpine rocks in the "Neues Jahrbuch" of the 1840s. One of the editors was Bronn, whom we have already met in connection with *Halobia* and *Monotis* (Chapter IV). People would write letters to Bronn. He would publish them and provide titles in the table of contents. In 1845 F.A.Quenstedt wrote from Tübingen that the St.Cassian and Hallstatt beds were Lower Cretaceous. Bronn evidently thought this was nonsense because he inserted an exclamation mark in his contents table ! Quenstedt was holding to this view as late as 1849. But Bronn himself also had the wrong answer. In 1847, reviewing Hauer's publication on the Metternich collection, he asserted that the Hallstatt beds were Jurassic. Quenstedt's mistake is commemorated in ammonoid nomenclature. He named a species from the Hallstatt Limestone *Ammonites neojurensis*. He chose the specific name to record his belief that it was Cretaceous, i.e. younger than Jurassic. Soon it would be shown by Hauer that it was Triassic. Today this ammonoid retains its name, as *Rhacophyllites neojurensis* (Quenstedt), a permanent reminder of Quenstedt's mistake. Quenstedt had fallen into the trap of trying to date an ammonoid on morphology alone, without knowledge of its position in the sequence. He was the authority on the Jurassic of Swabia, where the beds are displayed in orderly succession. Perhaps he should have stuck to his speciality !

All this[67] provides a measure for Hauer's achievements. It is mainly thanks to him that the Trias acquired faunal attributes that would permit its recognition around the world.

Now we may take stock of the status of the Trias at mid-19th century. Germanic Trias was named; the position in the sequence of the Muschelkalk ammonoid fauna, dominated by *Ceratites*, was known and appreciated. By 1856 Oppel and Suess had provided the clue for correlating the Triassic-Jurassic boundary between the Alpine and Extra-Alpine areas (Chapter II).[68] It was already clear that the Alpine faunas gave a much better record of marine life for the whole period. Index fossils for the Lower, Middle and Upper Triassic were known in sequence in the southern Alps. To name a few important ones: for the Lower, there was *Posidomomya clarae*; for the Middle, *Ceratites binodosus* and *Halobia lommeli*; for the Upper, *Ammonites aon*. These were of known significance and available for identification by mid-century. *Ceratites binodosus* was judged sufficiently close to the Germanic forms to warrant recognizing an Alpine Muschelkalk. In the northern Alps *Monotis salinaria* and a good variety of Hallstatt ammonoids were now known to be Triassic. Partly from position, partly by making the shrewd guesses that geologists have always made, there were reasonable grounds for supposing that Lower Alpine Triassic was about the same age as the Bunter; the Upper, the Keuper. Sequential data and described fossils were now available to start tracking the Lower, Middle and Upper Triassic over other parts of the globe. Now on to that.

ASIAN AND ARCTIC LOCALITIES

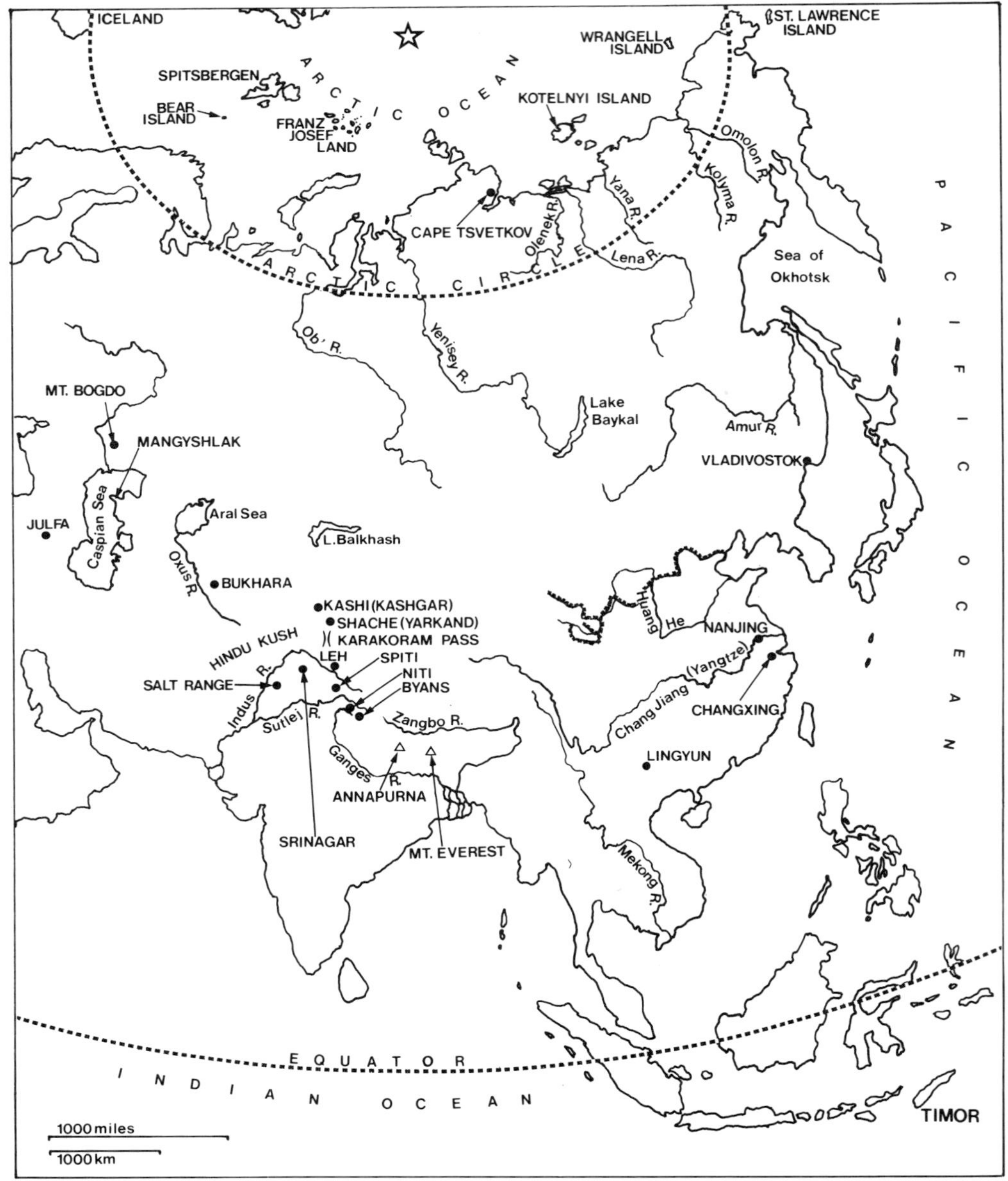

VI

Marine Trias found around the Arctic, Pacific and in Tethys

At the time that the Trias was being defined in Western Europe, geographic exploration was taking place in many parts of the world: the Arctic, the Himalayas, and in the lands bordering the Pacific. Some expeditions had geologists; most did not. But those without geologists often obtained specimens that were brought home to the scientists. In this way data accumulated and by 1860 it was clear that marine Triassic faunas occurred in widely scattered parts of the globe.

The very first collection was made around 1810, by Matthias Hedenstroem (1780-1845), on Kotelnyi Island, in the Arctic Ocean. Hedenstroem is believed to have been of Swedish birth.[69] He became an expatriate, living in St. Petersburg. Between 1809 and 1811 he was engaged by the Russian government to visit the New Siberian Islands (which include Kotelnyi) to seek information for settling disputes in the hunting monopoly. At the time these remote islands were of some economic importance, not only for hunting but also as a source of fossil mammoth ivory. Ivory, in the days before plastics, was a particularly useful and valuable commodity. The Russian government wanted to know more about the islands, and that is why Hedenstroem went there. He did not have an easy time. On his first trip he travelled with dog sledges; later with reindeer. On one occasion his party was saved from starvation by killing a dozen or so polar bears. His contribution to our story is that he found a piece of an ammonoid. This caused some interest in scientific circles, although not immediately. The discovery was announced in 1845 by the Russian professor, Eduard Ivanovich von Eichwald (1795-1876).[70] At about that time, in 1844, another expedition, under Alexander Theodor von Middendorf (1815-1894),[71] brought home Triassic fossils. Some were from the Olenek River, which reaches the Arctic Ocean near the mouth of the Lena, others from the Pacific shore of Siberia, on the sea of Okhotsk. The fossils collected by Middendorf, and the single specimen obtained by Hedenstroem, were described by Alexander Graf von Keyserling (1815-1891).[72] Keyserling was of a patrician Baltic family, born in the Duchy of Courland, then part of the Russian Empire. He studied under Humboldt and von Buch in Berlin. Sir Roderick Murchison was his friend and companion on geological travels throughout much of western Europe. In 1845 Keyserling described *Ceratites hedenstroemi*. In doing so he produced an interesting illustration, with the piece collected by Hedenstroem, from Kotelnyi Island fitted to another specimen collected some 600 km away, at the mouth of the Olenek, by Middendorf. Fair enough ! He explained what he had done, and the reader was not deceived. Keyserling concluded that the ammonoid suggested that rocks of Muschelkalk age were present in those distant parts. In 1848 Von Buch also showed interest and went even further; he regarded the Arctic ammonoid as a representative of a Muschelkalk species – *Ammonites semipartitus*. Von Buch's taxonomic conservatism is again evident; the genus *Ammonites*, not *Ceratites*, was used.

ARCTIC SQUADRON QUITTING THE NORE APRIL 1852

INTREPID **MONKEY** **ASSISTANCE** **AFRICAN** **PIONEER**

RESOLUTE **NORTH STAR**

This event led to the first discovery of Marine Trias in North America

From "The Last of the Arctic Voyages . . .", 1855[77]

Further discoveries in the Arctic came soon. In 1852 Sir Edward Belcher, C.B. (1799-1877),[73] discovered ichthyosaur bones on Exmouth Island, an islet in what is now the Canadian Arctic Archipelago. Belcher was born in Nova Scotia. Before going to the Arctic he did extensive marine survey work in the Pacific. It was for this, not his Arctic work, that he was knighted and created a C.B. It seems that he was not a very likeable person. Sir Clements Markham described him as "that notorious tartar" and judged him responsible for turning the hair of one of his officers prematurely grey.[74] In 1852 he was given command of an Arctic Expedition. He was provided with a squadron of five ships. This was the last of the British Admiralty Franklin Search Expeditions. Sir John Franklin (1786-1847) had sailed from England with two ships, the "Erebus" and "Terror", in 1845. His objective was to find a Northwest Passage. By 1852 there had been no news of the Franklin Expedition. The graves of three of Franklin's crew had been found at Beechey Island, a islet near the southwest corner of Devon Island. They had died in the winter of 1845-1846. This showed that "Erebus" and "Terror" had spent that winter in the ice nearby. What happened afterwards was unknown. By 1852 it was virtually certain that Franklin and his men had perished, but it was not known where, how or when. Belcher was instructed to find out. On entering the Archipelago in the summer of 1852, Belcher divided his squadron into three. Two ships ("Resolute" and "Intrepid") went west to Melville Island; two ("Assistance" and "Pioneer") north to the northwest corner of Devon Island; one, the "North Star", under the command of Captain W.J.S.Pullen, remained at Beechey Island. Past experience had shown that there was always open water at Beechey Island in the summer. "North Star" was thus a safeguard, in case the other ships became inextricably lodged in the ice. The waters around Beechey Island were not without hazards, however. On August 21, 1854 the "Breadalbane", a transport which had been sent out in support of the Belcher Expedition, was nipped in the ice and sank in 15 minutes, fortunately with no loss of life. The wreck has now been located and in May 1983 divers took photographs and recovered the wheel.[75]

"Resolute" and "Assistance" were bluff bowed barques, 410 and 430 tons, respectively, with polar bears as figureheads. "Pioneer" (ex "Eider") and "Intrepid" (ex "Free Trade") were smaller, sharp-bowed screw steamers, which acted as tenders to the larger sailing vessels.[76] Leopold M'Clintock, about whom more in Chapter XV, was Commander of "Intrepid". Belcher himself flew his flag on the "Assistance". The navigation season for "Assistance" and "Pioneer" ended on August 20, 1852. On August 23 Belcher set out on an exploratory journey to the north using a sledge and a small boat, the "Hamilton". On August 27 Belcher discovered and took possession of Exmouth Island. On the summit he found ichthyosaur bones that were eventually brought home and described by the eminent vertebrate paleontologist (and anti-evolutionist), Sir Richard Owen (1804-1892). Belcher gave a pretty good description of the geology of Exmouth Island, noting that the lower slopes were sandstone, and the summit limestone.[77] The sandstone is now known as the Bjorne Formation, the limestone, Schei Point. Because ichthyosaur bones are so common in the Jurassic of England, it was originally thought that the rocks on Exmouth Island were of that age.[78] Eventually, in 1957, Middle Triassic ammonoids were found at the same locality.

Belcher was not popular with his officers and was much reviled, because in the spring of 1854 he abandoned "Assistance" and "Pioneer" and gave orders that "Resolute" and "Intrepid" be treated in the same way. He intended to go home in the "North Star" and to take everybody with him. He saw no reason to share the fate of Franklin if it could be avoided. He was acting within his instructions but was nevertheless court-martialled on his return to England. He was judged not guilty, "but his sword was returned to him in a silence more eloquent than words".[79] He was never employed again. The sequel to this story is that the "Resolute" refused to accept abandonment. She became free from the ice and started

TRIAS DISCOVERY AND DISCOVERER

Sir Leopold M'Clintock
(1819 – 1907)

Sir Edward Belcher
(1799 – 1877)

The "Hamilton" under canvas. Exmouth Island is above the walrus. Table Island is in the distance. Both preserve fossiliferous Trias

From "The Last of the Arctic Voyages . . ." 1855[77]

back to England unattended ! She was recovered by American whalers on September 10, 1855, in Davis Strait, having drifted some 1600 km from where she had been left. The other three were presumably nipped and sank.

So the unpopular Sir Edward Belcher was the first to find marine Triassic fossils in the Western Hemisphere. His expedition found no trace of Franklin; this would be a later achievement, by Leopold M'Clintock (Chapter XV). But the expedition added a great deal to knowledge of the geography and geology of the Arctic Islands. It could be argued that some of the sledge parties explored in directions where it was virtually certain no trace of Franklin would be found. Most of Belcher's officers were more interested in making new geographic discoveries than in finding traces of their lost countrymen. Belcher went to Exmouth Island himself but was not a great traveller. However he took an interest in geology and examined collections brought back by his officers. Belcher's mineralogical expertise is revealed by the following quotations. G.H. Richards, Commander of the "Assistance", brought back rock specimens from Sabine Peninsula, Melville Island. Of the locality, Richards wrote "The lower land is sandstone; on the summit are large masses of lime, and I think some granite. Mica (at least so I take it to be) is very abundant in the cliffs. . . ". Belcher was not impressed by his officer's observations; he wrote "Pray number the intervening series between the shell-bearing limestone and granite. It is not for me to dispute these records; all have been told of their mistakes but determine to adhere to their adopted theories. The specimens are selenite".[80] The fact that Belcher's identification of selenite was correct, and of considerable significance, was not realized until 1953, when Bill Heywood of the Geological Survey proved that the circular structures in the Archipelago were piercement domes exposing selenite.[81] But from the tone of Belcher's critical remarks it is easy to believe that he was not a popular expedition Commander.

The next Arctic discovery was in Spitsbergen. Here the 1861 Expedition of Dr. Otto Torell (1828-1900) made collections, including ammonoids, described as *Ceratites* and recognized as Triassic, by Gustav Lindström (1829-1901)[82] in 1865. For many years he was in charge of the paleontological department of the Stockholm Rijksmuseum. This was a pretty good effort on Lindström's part, when one bears in mind that he was mainly concerned with much older – Ordovician and Silurian – faunas. The famous Swedish explorer and geologist, Baron Adolf Erik Nordenskiöld (1832-1901)[83] was a member of the 1861 Spitsbergen expedition and several later ones. In 1878-1879 he became the first to navigate the North-East Passage, in the "Vega". In 1867 Nordenskiold published an account of the Spitsbergen Trias noting that the Mesozoic rocks of Exmouth Island were almost certainly Triassic, not Jurassic, as had hitherto been assumed.[84] So Nordenskiold should be credited with the first recognition of marine Triassic in the Canadian Arctic. Belcher found the fossils; Owen described them, Nordenskiold dated them !

Discoveries were also made early in the Himalayas and Salt Range. Around 1825 Alexander Gerard brought back ammonoids from Spiti.[85] Spiti is in the western Himalayas and eventually became one of the most important localities for the marine Trias in the whole world. Ammonoids from Spiti had been known for centuries. Throughout India they were used as amulets, and in Hindu temples as fetish symbols. They are known as "salagrams".[86] Most or all salagrams are Jurassic or Cretaceous ammonoids, which are the most abundant in the Spiti area. Gerard, however, also brought back ammonoids that were correctly interpreted as Triassic by Henry Francis Blanford (1834-1893).[87] They included *Ammonites* (now *Ptychites*) *gerardi*, announced to the scientific world in 1863. H. F. Blanford and his brother William Thomas (1832-1905)[88] joined the Geological Survey of India in 1855. This organization was soon to have much to do with the marine Trias. The Blanford brothers had distinguished careers in geology and related scientific fields. Both became fellows of the Royal Society. The elder also received the Wollaston Medal in 1883, the year after Hauer (Chapter V). We meet W.T.Blanford again in Chapter IX.

CALIFORNIA AND INDIAN SURVEYS

CALIFORNIA SURVEY — ABOUT 1860
Back Row: W.M. Gabb, J.D. Whitney, C.R. King

GEOLOGICAL SURVEY OF INDIA — 1870
Standing: F. Stoliczka, R.B. Foote, W. Theobald, F.R. Mallet, V. Ball, W. Waagen, W.L. Willson.
sitting: A. Tween, W. King, T. Oldham, H.B. Medlicott, C.A. Hacket.

Another important figure in the early Himalayan studies was Richard Strachey (1817-1908), who eventually became General Sir Richard Strachey, G.C.S.I., Ll.D., F.R.S.[89] In 1848-49 Strachey, then a Captain in the Bengal Engineers, made traverses across the Himalayas. He must be credited with the earliest accurate interpretation of the geology. He recognized the great sedimentary terrane north of the metamorphic belt, and in it, rocks that he described as "Muschelkalk".[90] He collected a few ammonoids. They were sent to England for study by John William Salter (1820-1869),[91] the paleontologist for the Geological Survey of Great Britain. Strachey's ammonoids are described in a rare work – "The Palaeontology of Niti", jointly authored by Salter and H.F.Blanford, printed in Calcutta. It was intended to be an Appendix for a book on the Himalayas by Strachey. Regrettably the complete book was never published, but the Appendix appeared in 1865. But the people most concerned knew about the fossils and their significance, before then. In 1862 Suess was in London, his birthplace. He visited Salter, who showed him Strachey's ammonoids. Suess immediately recognized what he called *Ammonites floridus, Ammonites aon, and Halobia lommeli* – three important indicators for the Alpine Trias. His identifications were not exactly right, but were pretty good. They gave him a clear and correct message. Excited, Suess immediately wrote to Hauer in Vienna, telling him about the discovery. The letter was published in the "Verhandlungen" of the Geologischen Reichsanstalt.[92] The news was out ! Alpine Trias in the Himalayas. Another to grasp the significance of the Himalayan fossils at about this time was the Berlin Professor, August Heinrich Ernst Beyrich (1815-1896).[93] In 1866 he caught the significance of Blanford's *Ammonites gerardi*, correctly interpreting this ammonoid as characteristic of the Alpine Muschelkalk (i.e. Middle Triassic). This made possible another Alpine-Himalayan correlation.

Shortly after Strachey, the Schlagintweits made Himalayan collections (in 1854-1857). The three Schlagintweit brothers (Adolf, Hermann and Robert) made extensive explorations in the Himalayas and Tibet under the patronage of the East India Company.[94] Before going to Asia Adolf and Hermann had established themselves as accomplished alpinists, distinguished geographers and pioneer glaciologists. In 1850 they published a 600-page book on the physical geography, glaciers, meteorology and plant geography of the Alps. On September 15, 1851 they made the first ascent from the north of the Pyramide Vincent (4215 m), a peak in the Monte Rosa massif, on the Swiss-Italian border. In 1857 Adolf was "foully murdered (in Kashgar) by a scoundrelly robber named Wullee Khan". This account of Adolf's end was provided by the British explorer George Hayward, who crossed the Karakoram Pass to Kashgar in 1868-1869.[95] Wullee Khan later had his throat cut by Yakub Beg, whom we will meet in Chapter IX. Hayward also met a violent end (Chapter IX).

The Schlagintweits' fossils went to Munich and were described by Oppel and Gümbel. Oppel published on the ammonoids in 1863. Most of them are now known to be Middle Triassic but at that time Oppel had no grounds for making the perceptive analysis advanced by Suess and Beyrich. Oppel thought that most of them were Jurassic; one (now called *Ptychites impletus*) he presumed to be Triassic. A little later Gümbel wrote on the clams which he correctly determined to be Triassic.[96]

Another important Asian locality discovered in this period is the Salt Range, later to be known for its important section of Lower Triassic above Permian. In 1851, Dr. Alexander Fleming, Assistant Surgeon 4th Punjaub Cavalry, was there. Fleming had no great pretensions as a geologist. But he collected fossils and wrote letters to Sir Roderick Murchison, which were communicated to the Geological Society of London. The fossils were sent to the Belgian, Laurent Guillaume de Koninck (1809-1887).[97] Fleming's fossils included obvious Paleozoic things, like productid brachiopods, and also ammonoids with ceratitic suture lines. But Fleming could not provide much stratigraphy, and the fact that the Salt Range section revealed the Permian – Triassic boundary was not realized at this time.

Discoveries were also made on the western side of the Pacific in the early days. Around 1844 Middendorf's Siberian expedition found clams, described by Keyserling as "*Avicula ochotica*" on the shore of the Sea of Okhotsk, west of Kamchatka. Keyserling described these along with the ammonoids from the Arctic. He noted the similarites with Bronn's *Monotis*, but this was 1845, the age significance of *Monotis* was still uncertain, and the rocks were not dated as Triassic.

An important expedition was that of the Austrian Frigate "Novara", which sailed around the world in 1858-1859. On this there was a geologist – Ferdinand von Hochstetter (1829-1884),[98] who was born in Württemburg and died in Vienna. The ship called at New Zealand. Here, in January 1859, Hochstetter left the ship and then spent 9 months doing pioneer geological work on both North and South Islands. In the Nelson District (northern part of South Island) Hochstetter collected both *Monotis* and *Halobia*. These were described by Zittel in 1864 and, correctly interpreted as Triassic. Triassic on the western side of the Pacific was firmly established. Discoveries in Japan and near Vladivostock (Primory'e) came later. Edmund Naumann (1854-1927)[99] announced the discovery of *Monotis* in Japan in 1881. Naumann was one of the Europeans brought to Japan during the modernisation period, which started in the 1860's, following the initiative of both the United States and Russia, to open up diplomatic relations, trade and communications between Japan and the western world. Commodore Perry's visit, it will be recalled, was in 1853. Naumann's job was to direct the Geological Survey.

Discoveries in western North America start around 1860, with the formation of the Geological Survey of California, under the leadership of Josiah Dwight Whitney (1818-1896).[100] He was not titled, like the European patricians, but patrician nonetheless. The work of this Survey led to the discovery of *Monotis* at Taylorsville, California; also in what is now Nevada. Nevada was not then a State in the Union; this did not come until 1864. California was pretty remote in those days. In the East the Civil War was raging. Not until 1869 was there a transcontinental railway. The state government was not wholly supportive of Whitney's efforts in pure geology. They would have preferred him to find more of the gold that not so long before had drawn the Forty-Niners. At one stage Whitney had to borrow money from his brother to keep things going, but he was eventually repaid by the State. Whitney formed a staff which included William More Gabb as paleontologist. Born in 1839, Gabb died in 1878 of tuberculosis.[101] In 1864 Gabb described *Monotis subcircularis* from the California locality and also an assortment of ammonoids from the Humboldt and East Ranges in present-day Nevada. All these he recognized as Triassic. Most of the Nevada specimens were donated to Whitney by friends – Gorham Blake Esq., and R. Homfray; they were not collected by his staff. The *Monotis* locality in California was found by a staff member – Clarence R. King (1842-1901).[102] A few years later King was to have his own Survey: that of the 40th Parallel funded by the Federal Government (1867). This was one of the "Geological Surveys of the Territories" undertaken for about a decade before the formation of the United States Geological Survey. Eventually King became the first Director of the United States Survey, when it was formed in 1879. Another of the Territorial Surveys was that of the Upper Missouri Country led by Ferdinand Vandiveer Hayden (1829-1887).[103] The King Survey found new Triassic localities in Nevada and the Hayden Survey led to the discovery of the Lower Trias of the eastern part of the Cordillera. All this led to published results by Fielding Bradford Meek (1817-1876),[104] in 1877, and Charles Abiathar White (1826-1910),in 1880. White was trained as a medical man but later took up paleontology.[105] The discovery of smooth ceratitic ammonoids – *Meekoceras* – dates from this period. These were soon to assume worldwide importance for recognizing the Lower Triassic.

First discoveries in the northern part of the Cordillera came between 1870 and 1887. A

PIONEER CANADIAN WORKERS

A.R.C. Selwyn
(1824 – 1902)

G.M. Dawson
(1849 – 1901)

R.G. McConnell
(1857 – 1942)

J.F. Whiteaves
(1835 – 1909)

PEACE RIVER, BRITISH COLUMBIA

Peace River, looking upstream to Ne-Parle-Pas Rapids. The drawing was published in the Geological Survey of Canada Report of Progress for 1875-1876, from a photograph taken by A.R.C. Selwyn, 21 September, 1875.

discovery in Alaska was first of all. This was made by a French expedition (1870-1872) under Alphonse Pinart (b. 1852). Specimens of *Monotis* were found at Puale Bay, and recognized as Triassic by the eminent Parisian malacologist and paleontologist Paul Henri Fischer (1835-1893).[106]

Discoveries in British Columbia by pioneers of the Geological Survey of Canada followed soon. The Survey had been founded in 1842 with William Edmund Logan (1798-1875) as its first Director. This was before Confederation, so the Survey, in its early years, was concerned only with the east, where there is no marine Trias. The second Director was Alfred Richard Cecil Selwyn (1824-1902).[107] Selwyn was born in England. After a period with the British Survey he became Director of the Survey of Victoria Colony, Australia. In 1869 he was appointed Director of the Canadian Survey, a position he held for 26 years. This was on the recommendation of Logan, who by then was Sir William. He had been knighted by Queen Victoria at Windsor Castle in 1856. Not surprisingly the appointment of an outsider caused some dismay among some of Logan's staff, notably T. Sterry Hunt (1826-1892). Hunt, a notable and controversial figure in late 19th century geology, will make another appearance in Chapter VII. Selwyn's years as Director must have been difficult. He was harassed by his staff and also by committees of Parliament. The Survey had a very high profile in those days and Selwyn was often in direct communication with the Prime Minister. These difficulties did not turn Selwyn into a desk-bound administrator. He was an active field worker in many parts of Canada. In 1871 British Columbia became a part of Canada, and the responsibilities of the Survey were enlarged to include the whole of the Canadian Cordillera. Selwyn himself took part in this work and in the summer of 1875 travelled from the Pacific Coast, north through the interior of British Columbia, and crossed the Rocky Mountains by way of Peace River. In the fall he returned by the same route. On this journey he became the first to prove Trias in the Canadian Cordillera, by finding *Monotis* west of Ne-Parle-Pas Rapids on the Peace. Selwyn was accompanied on the east-bound journey by John Macoun (1831-1920), the Irish born, self-educated botanist and ornithologist who was the first to appreciate the agricultural potential of Canada's western plains. Although not a geologist Macoun became a Survey employee, ranked as an Assistant Director under Selwyn's regime.[107] The explanation for this is that starting in 1877 and for some years following, the organization was known as the "Geological and Natural History Survey of Canada", with extra responsibilities that nowadays are dispersed to the National Museums and other organizations. Macoun had taken the Peace River route in 1872 and it was for this reason that Selwyn was determined to have him on the 1875 expedition. Macoun's 1872 journey had been with Charles Horetzky (b. 1838). Both were members of Sanford Fleming's expedition seeking a route through the Rockies for the Canadian Pacific Railway. Fleming delegated Horetzky and Macoun to investigate the northern routes. Fleming himself examined the passes in the southern Rockies that were eventually used, leading to the completion of the Canadian Pacific Railway in 1885. Horetzky made an unsuccessful attempt to cross the Rockies at Pine Pass, south of the Peace. Macoun, meanwhile, had started for the Peace. Horetzky later joined Macoun and together they took the Peace River route and went on to the Pacific coast. Peace River never got a railway through the mountains, nor even a road. Eventually there would be a dam and a lake (Chapter XIV). Pine Pass eventually got both, a road in about 1950 and a railway in 1958. This led to important exposures of *Monotis* beds in both road and railway cuts which became well known, in 1962, from the work of G.E.G. Westermann.

Selwyn's successor as Director, George Mercer Dawson (1849-1901), made extensive observations on the Triassic of British Columbia. Dawson was the courageous, brilliant pioneer who worked throughout most of Western Canada. He was the son of Sir William Dawson (1820-1899), the Principal of McGill University in Montreal, well known for his

FIRST AERIAL PHOTOGRAPHY

FIRST AERIAL PHOTOGRAPHY OF THE GRAND CANYON OF LIARD RIVER, AUGUST 2, 1935. The Fairchild 71 aircraft and crew: Bill Sunderland, photographer; C.H. ("Punch") Dickins, pilot; A.D. "Dan" McLean (1896-1969), Superintendent of the Civil Aviation Branch, Department of Transport; Charles Camsell (1876-1958), Deputy Minister of Mines. Photographs were taken obliquely from an elevation of about 2,500 feet. The camera is on the dock. Above: the narrow, totally unnavigable canyon where an anticline of hard Permian rocks intersects the river's course. The Devil's Portage is to the left (south) of the canyon. Opposite, above: the Rapids of the Drowned; Lower and Middle Triassic rocks form the canyon walls. Opposite, below: Hell Gate, the lowest of the succession of narrow canyons exposing hard Triassic rocks. Between the canyons are stretches where the river is wide, with placid water. Here the rocks are soft Cretaceous shale. The distant canyon is where R.G. McConnell collected *Nathorstites* in 1887.

LIARD RIVER

pioneer geological work in eastern Canada. When 11 years old Dawson junior contracted an illness which left him a stunted hunch-back; he never attained a height of 5 feet. At the time the diagnosis was vague, but it seems likely that he suffered an attack of polio. In later years he would be known to his colleagues as "The Little Doctor"; to his Indian field companions as "Skookum Tumtum" – "strong, tough man . . . brave cheery man".[108] Undeterred by his physical handicap he went to the Royal School of Mines in London for training and then embarked on a career that took him throughout western Canada. In 1873 he was appointed to the British North America Boundary Commission and in 1875 joined the Geological Survey of Canada. In 1878 he collected *Monotis* and other Triassic fossils in the Queen Charlotte Islands, and at other times between 1875 and 1894 made important observations and collections from the Triassic of southern British Columbia (Nicola Group), and Vancouver Island (Vancouver Group). In 1887 he noted the occurrence of limestones on the east side of Lake Laberge, in the Yukon. He collected corals and other fossils but their significance was not clear, and it was not until Everett J.Lees went there in 1929-1930 that it became known that the rocks were Triassic.[109]

The journey of Richard George McConnell (1857-1942)[110] that led to Trias discoveries was an epic in the history of the Geological Survey of Canada. McConnell was a member of the Yukon Exploring Expedition of 1887-1888, led by G.M.Dawson.[111] They travelled together from Wrangell, Alaska, up Stikine River, crossing the continental divide to Dease River. They parted where the Dease joins the Liard. Dawson went up the Liard and made a broad sweep through the Yukon, which eventually brought him to Lake Laberge and the Pacific coast, at the Chilkoot Pass. McConnell went downstream on the Liard. Liard River traverses a belt of Triassic rocks, some 50 km wide. Much of the rock is resistant sandstone, forming narrow canyons, known as "gates", where the river is swift and turbulent. Experienced boatmen are known to have navigated all the Triassic gates.[112] Hard Permian chert exposed on one bend forms a canyon where the river is totally unnavigable. Here it is necessary to portage over a ridge that rises nearly 300 m above the river. This was appropriately named the Devil's Portage. Below the portage is the Grand Canyon of the Liard. This has been run many times but not without taking its toll. One stretch is known as the "Rapids of the Drowned" in memory of a clerk of the Hudson's Bay Company named Brown, whose boatload of voyageurs upset, with total loss of life. "Hell Gate" marks the lower end of the Grand Canyon; to the west-bound voyageurs, going upstream, this marked the start of the bad water. This stretch of the Liard was McConnell's route, soon after parting from Dawson. He left with two white companions and two Indians, but the Indians soon deserted. They were using a small wooden boat they had built at Dease Lake. When they reached the Devil's Portage the boat proved too heavy for the route. Six days were then taken to make the arduous portage and construct a canvas boat on a wooden frame. The canvas, sewn in the shape of a boat, had been brought from Ottawa for the purpose. It was painted with half a gallon of oil but would not float until also coated with a sticky mess of sperm candles, gun oil and bacon grease stirred up with spruce gum. On July 16, 1887 the boat was launched and they started down the Grand Canyon. Their faith in the boat was limited, however, and they made several portages before reaching the relatively serene river at Hell Gate. Triassic fossils were collected at several localities in the Grand Canyon, including the ammonoid now known as *Nathorstites*, which was found at the first canyon above Hell Gate. This genus would turn up later on Bear Island and at many other places in the Arctic. A handsome ammonoid collected by McConnell eventually became the type species of the genus *Dawsonites*. Sad to say the name had already been used for a fossil insect. Under the rules the ammonoid name had to be changed – to *Daxatina*. So, regrettably, G.M.Dawson's name is no longer commemorated in Triassic ammonoid nomenclature.

After traversing the Triassic belt McConnell continued downstream on the Liard. Above the forks with Fort Nelson River he met Hudson's Bay Company voyageurs under W. Lepine making their way upstream. McConnell decided to send his two white companions back with Lepine. He went on to Fort Liard, for a while with an Indian, but mostly alone. At Fort Liard he obtained a bark canoe and then carried on to where the Liard joins the Mackenzie at Fort Simpson. He then spent the winter of 1887-1888 at Fort Providence, on the west end of Great Slave Lake. In the summer of 1888 he descended the Mackenzie to Peel River; thence via the portage to Porcupine and Yukon rivers, finally up the Yukon to Chilkoot Pass and the Pacific Coast. In all he had travelled about 6400 km, 1600 by foot, 4800 by water, making track surveys nearly all the way. He had made two complete traverses of the Canadian Cordillera and had crossed the Interior Plains as far as the westernmost exposures of the Canadian Shield, on Great Slave Lake. McConnell eventually became Deputy Minister of Mines (1914-1920).

The fossils collected by Selwyn, Dawson and McConnell were described by John Joseph Frederick Whiteaves (1835-1909),[113] paleontologist, zoologist and museum curator for the Geological Survey. Whiteaves and Macoun became colleagues during the Selwyn regime. It cannot have been a happy relationship. Macoun gained his position, as Assistant Director and Naturalist though the intervention of the Minister of the Interior, without Selwyn's knowledge or approval. As such he claimed administrative superiority over Whiteaves, a relationship that Selwyn, and presumably also Whiteaves, were reluctant to admit. The devastation of the National collection of bird skins did not help matters ! Whiteaves, as curator, was presumably responsible for them and during his tenure they were mostly destroyed by moths and dermestid beetles. Evidently he was more attentive of the relatively indestructible fossils. Many of skins lost had been collected by Macoun.

Finally: South America. *Monotis* and Triassic ammonoids were found in Peru in 1875. There's a funny story here. Gustav Steinmann, the erratic genius we've already met, became much involved in Peruvian studies. The occurrence of Triassic fossils in Peru had been mentioned in publications as early as 1886 by Mojsisovics. Early in 1909, Steinmann writes in the German periodical – "Neues Jahrbuch" – that there is NO TRIASSIC IN SOUTH AMERICA; those who have said so are WRONG. Within the year, in the same journal, he writes that there IS Triassic in South America.[114] Why he chose to knock the earlier reports I cannot understand. But he seems to have been a perverse and argumentative fellow. We have already seen (Chapter III) that he got into great arguments with Diener about ammonoid phylogeny.

Here we will end the review of mid to latish 19th Century discoveries in parts of the world remote from the typical Trias. What had been achieved ? I would say a lot, in terms of indicating the worldwide extent of marine Triassic rocks. The pattern: circum-Arctic, circum-Pacific and Tethyan (but not circum-Atlantic) was more or less clear. This significant generalisation holds true today; significant because it gives more than a clue that there was no Atlantic in the Trias. But the Trias was still pretty gross; finer divisions were not very clear, although in many places it was possible to assign rocks to one or other of three divisions: Lower, Middle and Upper Triassic. Now we move on to the efforts and achievements towards refining the time scale.

ALPINE AND HIMALAYAN WORKERS

E. von Mojsisovics (1839 – 1907)

János Böckh
(1840 – 1909)

C.L. Griesbach
(1847 – 1907)

VII

1867-1892: A Quarter Century of Mojsisovics

Around 1867 the important figure of Mojsisovics entered the field of Trias studies. Johann August Georg Edmund Mojsisovics, Edlen von Mojsvar was born in Vienna, October 18, 1839, of an old Hungarian family.[115] At first he studied law but he liked the outdoors and mountains and in 1865 joined the Geologischen Reichsanstalt as a volunteer. By 1867, when von Hauer was Director, he was an official member; from 1873 a Chief Geologist; between 1893 and 1900 the Vice-Director. In 1900 he retired with the honorary title of "Hofrath". He left the position of Vice-Director before his time was up. His loss of position was related to the bitter dispute that developed between Mojsisovics and a Reichsanstalt colleague – Alexander Bittner. This dispute has a chapter of its own (VIII).

Mojsisovics was in the Vienna scientific establishment. He became a corresponding member of the Imperial Academy in 1883, and a full member in 1891. He died in 1907. He is known for his field studies in both the northern and southern Alps and for his monographs on Triassic ammonoids from all over the world – the Alps, Arctic and Himalayas in particular. His descriptive monographs on Triassic ammonoids are gigantic quarto volumes, mostly published by the Reichsanstalt. His biggest single monograph, on the Upper Triassic ammonoids from the Hallstatt Limestone, published between 1873 and 1902, is illustrated by 223 lithographic plates. Another, "Die Cephalopoden der Mediterranen Triasprovinz", which appeared in 1882, has 94. They are weighty contributions. My own set weighs 20 kilos! The 1882 monograph has descriptions of Lower, Middle and Upper Triassic ammonoids from Hungary as well as Austria, also from south Tirol, then a part of Austria, now in Italy. Most of the Hungarian ammonoids were from the hills north of Lake Balaton and had been originally described by Janos Böckh de Nagysur (1840-1909),[116] who from 1882 until his death was Director of the Hungarian Geological Institute in Budapest. Many of the specimens illustrated by Mojsisovics can be seen today in the Vienna and Budapest Museums and compared with the illustrations. Nearly all the illustrations were engraved directly on the stone and as printed are reversed. With most ammonoids this does not matter because they are coiled in a plane spiral. For *Cochloceras*, coiled like a screw or a snail, the lithographer used a mirror, so that the printed picture would look right. Some of the illustrations are restorations, with missing pieces filled in, but nearly all are scientifically accurate and useful. The detailed accurate descriptions and illustrations provided by Mojsisovics are unquestionably the greatest contribution by a single author towards appreciating the astonishing beauty and variety of Triassic ammonoids.

Mojsisovics was also very interested in Triassic biochronology. Here he did not do so well. I have the impression that he wanted to divide the Triassic in the way that Oppel had the Jurassic, some 20 years before. When Oppel divided the Jurassic into 33 zones, he did it from stratigraphy. Mojsisovics tried to do something comparable for the Trias. He had

SALZKAMMERGUT TRIAS – I

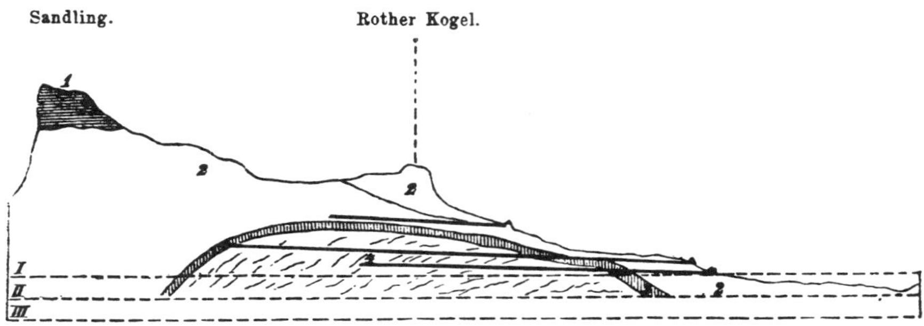

Above: Interpretation of the geology east of Sandling Mountain given by Hauer in 1875. 1, Jurassic; 2, Hallstatt Limestone; 3, Zlambach Marl; 4, Salt deposits. Below: view across the Leisling valley. Sandling Mountain is to the right. The rocky wall is the Leislingwand, exposing Hallstatt Limestone.

plenty of ammonoids, but very little stratigraphy. He made many mistakes by producing tables showing faunas in sequence when the sequence had not been observed.

He got under way in 1869 by defining the Carnian and Norian stages.[117] Carnian was typified by beds with *Ammonites aonoides*. Norian included beds with *Ammonites metternichi*. The boundary was said to be within the Hallstatt Limestone. The Carnian fauna is in the Hallstatt Limestone; the Norian fauna is in both the Hallstatt and the Zlambach Marl. Both these formations occur in Salzkammergut, where the stratigraphy is difficult and the structure very complicated. The geology is not easy to interpret. Even today there is not unanimous agreement about the relationships. The best exposures of the Zlambach Marl are in the valleys. Exposures of the Hallstatt Limestone occur on the surrounding hills and mountains. Well past the middle of the 19th Century geologists supposed that the Hallstatt Limestone, being higher topographically, was stratigraphically higher than the Marl. Nowadays it is believed that the Marl is partly above, and partly equivalent to the Limestone. But it is not obvious. So I don't think Mojsisovics should be blamed for thinking that the Zlambach Marl was below the Limestone. Hauer and Stur thought the same.[118] The trouble was, this supposed stratigraphy led to the belief that the Norian was older than the Carnian.

Two years later Norian was proposed as a geological term with an entirely different meaning by Thomas Sterry Hunt, one of Logan's workers in the early days of the Geological Survey of Canada. This was in his address as ex-president of the American Association for the Advancement of Science, given at Indianapolis in August 1871. Hunt's Norian was for Precambrian rocks which he believed to be younger than Logan's Laurentian, which at that time was considered to include the oldest of all known rocks. Precambrian Norian was proposed from occurrences in Canada, but the name is derived from Norway. Triassic Norian is from the Roman Province Noricum. So from both the etymological and geological standpoint the two could not be more different ! Hunt's ideas about Precambrian history are now outdated and Precambrian Norian has been nearly forgotten.

In addition to working in Salzkammergut, Mojsisovics was active in the Dolomites: the St. Cassian district and its environs. Here, as we have seen, the stratigraphy is clearly displayed. There are quite a few ammonoids, but most of the Hallstatt and Zlambach species are not present. There were two notable links, however, one permitting correlation in the Middle Triassic; the other enabling at least a rough correlation between the St.Cassian beds (with *Ammonites aon*) and typical Carnian – the *Ammonites aonoides* beds of the Hallstatt Limestone. Confronted with the absence of most of the Hallstatt ammonoids in the southern Alps, Mojsisovics decided, in 1874, that he was dealing with two separate faunal provinces.[119] The northern province, where most of the Hallstatt ammonoids had lived, became the Juvavian Province; the southern, the Mediterranean Province.

All along, Mojsisovics was writing extensively. In 1873 he launched Monograph 6 of the Reichsanstalt, with the grand overall title: "Das Gebirge um Hallstatt" – the Mountains around Hallstatt. He intended this to be a monograph on both the geology and paleontology of the Hallstatt and Zlambach beds. In that sense it was never finished. The scope contracted and the later parts were titled "Die Cephalopoden der Hallstätter Kalke".[120] In this monograph Mojsisovics dealt with all the collections that he could lay his hands on, including those previously described by Hauer and Dittmar. Most of the specimens that Hauer had described were in Vienna and available to Mojsisovics. As mentioned in Chapter V, the Metternich collection, the subject of Hauer's first paper on Triassic ammonoids, seems to have been an exception. From the Berlin Museum Mojsisovics borrowed an important collection which had been described by Dittmar.

Alphons Julievitch von Dittmar published a major work on Hallstatt ammonoids in 1866. His work thus forms a link between that of Hauer and Mojsisovics. It was based on collections made by Dr.von Fischer of Munich. Fischer was physician to Maximilian II,

SALZKAMMERGUT TRIAS – II

Above: Alexander and Edith Tollmann at the south side of the Feuerkogel, 1966. The outcrop is Hallstatt Limestone, of Carnian and Norian age. Below: southerly view over Hallstatt Lake. The glacier is on the Dachstein. The town of Hallstatt is on the small delta. Salt mines, Sommeraukogel and Steinbergkogel are in the cleft above the town, to the right.

King of Bavaria. He seems to have been an avid fossil collector because a few years earlier he had provided specimens that were described by Hauer and Moritz Hoernes. Relatively little seems to be known about Dittmar. He died near Riga in 1903. In the early 1860s his scientific activities were centered in western Europe. In Chapter II we saw that in 1864 he published a book on the Contorta-Zone, the Triassic-Jurassic boundary beds. From 1867 he worked mainly or wholly in Russia. Some of the Hallstatt ammonoids that he described are in the Humboldt Museum, Berlin, but most of those that were subsequently described by Mojsisovics seem to be lost. Dittmar's work includes the description of many new ammonoids. He also gave a complete review of the earlier work and compared the Hallstatt faunas with those from 28 other Alpine localities. Dittmar seems to have been the first to attempt to arrange the Hallstatt faunas in a chronological sequence. There was not much stratigraphic information available on which to base his ideas. Dittmar made use of what had been written but apparently did not make any observations of his own. I can find no record that he collected Hallstatt fossils himself, or indeed even went to Salzkammergut. His main source of sequential data is from an account given by Suess on the supposed succession north of the Leisling Valley. This area is west of Sandling Mountain, and includes the famous Millibrunnkogel locality. Suess' account was given in a paper published by Hauer in 1860. Suess distinguished several beds; these he seems to interpret as forming an upward succession overlying the salt deposits. Dittmar, in his own publication, took Suess' section and turned it upside down, with the salt deposits at the top. In making this change Dittmar states that Suess also thought that the sequence was upside down, but I cannot discern this in Suess' account. Dittmar focussed attention on two important beds described by Suess, the "Fasselschicht" (which he designated "a") and the "Gasteropodenschichten" ("b"). Fasselschicht was a name given by the local collectors. It means "Barrel Bed" in allusion to the abundant occurrence of a barrel-shaped ammonoid, now known as *Tropites subbullatus*. Gasteropodenshichten means snail beds; besides snails these beds also provided numerous ammonoids. In today's terminology the fauna of "a" is Carnian, of "b", Norian. Dittmar thought that "a" underlay "b"; in other words that Carnian was older than Norian. He also saw similarities between the fauna of "b" and some of the other Hallstatt faunas now recognized as Norian. These were the right answers. Dittmar's conclusions were not expressed in this terminology because Carnian and Norian had not been named in his day. This did not come until 1869, with the paper by Mojsisovics in which Carnian was supposed to be younger than Norian. It seems strange, but Dittmar's views are not referred to in Mojsisovics' paper. They also seem to have been overlooked in the 1890s, when the controversy over the Norian was raging (Chapter VIII). Although Dittmar was basically right concerning the age relationship of the faunas, it was not wholly for the right reason. The stratigraphy and structure of the area described by Suess is now known to be far more complicated that anybody thought in the 1860s. I am not aware of any place where simple stratigraphic juxtaposition of Carnian and Norian is demonstrable, whether right side up or upside down.[121] There will be more about this in Chapter XVI.

In 1879 Mojsisovics published " Die Dolomitriffe von Sudtirol . . . ."[122] This is a fine book on the southern Alps, often acclaimed as a classic in the study of fossil reefs. I am not a reef expert but I sometimes wonder if Mojsisovics' contribution in the field is overrated to the detriment of his predecessor, Ferdinand Freiherr von Richthofen (1833-1905), who published a pioneering work on the same subject and area in 1860.[123] Richthofen was born and died in Germany. The Red Baron was a relation.[124] Richthofen travelled extensively in China and North America. Around 1860 he was in California, where he met Whitney. The two became life-long friends.[125] Maria Ogilvie (later Ogilvie-Gordon) was also a friend (Chapter XII).

In his books and other publications Mojsisovics proposed biochronological schemes for the Triassic. Nearly all were different. All have zones in the wrong sequence. As science

TABLE III. TRIASSIC TIME SCALE – 1895

Serien	Stufen	Unterstufen	Mediterrane Triasprovinz: Zone (der pelagischen Facies)	Mediterrane Triasprovinz: Schichtbezeichnung (verschiedenartiger örtlicher Entwicklung)		Indische Triasprovinz: Zone (der pelagischen Facies)	Indische Triasprovinz: Schichtbezeichnung (verschiedenartiger örtlicher Entwicklung)
Bajuvarisch	Rhaetisch		22. Z. der *Avicula contorta*	Koessener Sch.	Dachsteinkalk		
	Juvavisch	oberjuvavisch (Sevatisch)	21. Z. des *Sirenites Argonautae*	Juvavische Hallstätter Kalke		Juvavische Cephalopodenfaunen des Himalaya	
			20. Z. des *Pinacoceras Metternichi*				
		mitteljuvavisch (Alaunisch)	19. Z. des *Cyrtopleurites bicrenatus*				
		unterjuvavisch (Lacisch)	18. Z. des *Cladiscites ruber*				
			17. Z. des *Sagenites Giebeli*				
Tirolisch	Karnisch	oberkarnisch (Tuvalisch)	16. Z. des *Tropites subbullatus*	Sandling Sch.		Karnische Cephalopodenfaunen des Himalaya	
		mittelkarnisch (Julisch)	15. Z. des *Trachyceras Aonoides*	Raibler Sch.			
		unterkarnisch (Cordevolisch)	14. Z. des *Trachyceras Aon.*	Cassianer Sch.			
	Norisch	obernorisch (Longobardisch)	13. Z. des *Protrachyceras Archelaus*	Wengener Sch.			
		unternorisch (Fassanisch)	12. Z. des *Dinarites avisianus*	Marmolatakalk			
			11. Z. des *Protrachyceras Curionii*	Buchensteiner Sch.			
Dinarisch	Anisisch	Bosnisch	10. Z. des *Ceratites trinodosus*	Oberer Muschelkalk		Z. des *Ptychites rugifer*	Muschelkalk des Himalaya
		Balatonisch	9. Z. des *Ceratites binodosus*	Unterer Muschelkalk		Z. des *Sibirites Prahlada*	Brachiopoden-Schichten mit *Rhynchonella Griesbachi* (Himalaya)
	Hydaspisch					8. Z. des *Stephanites superbus*	Obere Ceratiten-Kalke der Salt Range
Skythisch	Jakutisch		Z. des *Tirolites Cassianus*	Werfner Schichten der Ostalpen		7. Z. des *Flemingites Flemingianus*	Ceratiten-Sandstein der Salt Range Subrobustus Beds des Himalaya
						6. Z. des *Flemingites radiatus*	
						5. Z. des *Ceratites normalis*	
	Brahmanisch	Gandarisch				4. Z. des *Proptychites trilobatus*	Ceratite Marls der Salt Range
						3. Z. des *Proptychites Lawrencianus*	
						2. Z. des *Gyronites frequens*	Untere Ceratiten-Kalke der Salt Range
		Gangetisch				1. Z. des *Otoceras Woodwardi*	Otoceras Beds des Himalaya

The interpretation of Mojsisovics, Waagen and Diener.[127]

develops we all make mistakes. For making mistakes, there is no reason to criticise Mojsisovics. But he was naughty to say that faunas occurred in sequence without giving evidence. The reason he did not, it eventually transpired, was that he had almost none (Chapter XII).

In 1893, Mojsisovics produced a zonal sequence for the whole of the Alpine Trias.[126] The 1893 sequence, with some minor changes, appeared again in a more comprehensive table published in 1895 (Table III).[127]

Let us consider the vicissitudes experienced, between 1869 and 1893, of some of the younger zones (15-21 of Table III). It is a complicated history ! To start with, remember that 15 (Aonoides) is typical Carnian; 20 (Metternichi), typical Norian. As he found new zones, Mojsisovics was entitled, using the hierarchial system (Chapter II), to regard the new ones as Carnian or as Norian. But Carnian must forever include 15, and Norian 20. In 1869 the sequence was 20, 15;[117] in 1874, 20, 16, 15;[119] later in 1874, 20, 17(= 18), 19, 16, 15;[128] in 1879, 20, 19, 18, 19, 16, 15;[122] in 1892, 15, 16, 17, 18, 20, 19, with a few others thrown in;[129] in 1893, 15, 16, 17, 18, 19, 20, 21.[126] Don't worry if this seems hard to follow ! The object is to show that he produced several schemes in which the same zones appeared in different sequence. Also, in some schemes, zones that eventually turned out to be the same age, were arranged as if in sequence.The most important thing is that between 1869 and 1879 Norian was below Carnian; from 1892 on, above. Zone 21 (*Sirenites argonautae*) was introduced for the first time in 1893 and accordingly does not figure in the earlier schemes. This is the only Salzkammergut zone proposed for beds not in the Hallstatt and or Zlambach formations. The 1895 table is misleading in this respect but from his writings, Mojsisovics' intention is clear. He had the Argonautae Zone only in the Pötschen Limestone, a different Salzkammergut formation. Mojsisovics believed that the Pötschen limestone was very high in the sequence, but again, no evidence was given. Putting the Argonautae Zone so high was to cause confusion for nearly 100 years ! We now know it to be Columbianus Zone (Table II); or Hogarti Zone, to use a more precise division recently introduced for the earlier part of the Columbianus Zone.[130] For the interval we have considered, Mojsisovics thus produced at least four different schemes between 1869 and 1893. Regrettably all were wrong.

By 1892 Mojsisovics had decided that in Salkzammergut the Norian was younger, not older, than the Carnian. On this point he was, at last, right. I don't know why Mojsisovics changed his mind. He gave no stratigraphic evidence to support the original, wrong interpretation. But nor did he provide evidence for the correct one !

In 1892 Mojsisovics also decided to abandon the distinction between Juvavian and Mediterranean faunal provinces. He was finding that Juvavian ammonoids were popping up all over the world. Having changed his mind about both the succession and the faunal provinces, Mojsisovics came up with a new nomenclature. For the original Norian rocks (characterized by *Ammonites metternichi*) he adopted his old provincial name, Juvavian. Beds in the southern Alps, which truly underly Carnian strata, he called Norian.[129] For this he had some excuse, because in the 1869 paper[117] he had assigned some truly pre-Carnian beds to the Norian, in addition to those with *Ammonites metternichi*. This was nevertheless a cover-up. In his tables, Norian remained below Carnian – the relationship he originally thought prevailed in Salzkammergut. But Mojsisovics' Norian beds, in this scheme, are much older than the original Salzkammergut Norian with *Ammonites metternichi*. Now, undeniably, there were two Triassic Norians. By proposing this scheme Mojsisovics made a big mistake; one that was to cause heated argument and polemics, with polarization and division of the Vienna geological community for a whole decade. The cover-up was soon spotted by Mojsisovics' younger colleague in the Reichsanstalt – Alexander Bittner. The war was on. In order to follow the battles that followed I must label the two Norians. The younger, original one will be the "Metternichi Norian"; the older, found in the Dolomites, the "Archelaus Norian".

EUROPEAN WORKERS

G.G. Gemmellaro
(1832 – 1904)

Carl Diener
(1862 – 1928)

Alexander Bittner
(1850 – 1902)

Fritz Frech
(1861 – 1917)

G. von Arthaber
(1864 – 1943)

VIII

1892-1902: A Decade of Controversy Bittner versus Mojsisovics

Now is the time to introduce Alexander Bittner.[131] Bittner was born in Bohemia, March 16, 1850. Early schooling was there; later he went to the University in Vienna. Between 1877 and his death, in 1902, he worked for the Geologischen Reichsanstalt. He was thus a colleague of Mojsisovics. Dr. Franz Tatzreiter, who until recently worked for the Bundesanstalt (the successor to the Reichsanstalt), tells me that Mojsisovics and Bittner were not only colleagues but actually shared the same office. Considering what was to about to happen, in the 1890s, it seems likely that the atmosphere in this office may have become tense, to say the least.

Bittner was a field geologist and paleontologist, his speciality being clams and brachiopods. He seems to have been greatly respected by most of his Reichsanstalt colleagues, but this may not say very much for them. Bittner was a spartan individual who devoted all his energy to work; a bachelor, he lived with his sister. He was asthmatic and on Easter Sunday 1902 nearly suffocated. But he recovered and was thought by the doctor to be out of danger. The prognosis proved overly optimistic: he died on Easter Monday.

Mojsisovics, we have seen, had training in law; Bittner did not. But when it came to laws of priority in stratigraphic nomenclature, Mojsisovics was to prove no match against Bittner.

We have seen in the preceding Chapter that in 1892 Mojsisovics proposed to use the geological term "Norian" differently, compared with his original definition. He was advocating Archelaus Norian; and suppressing Metternichi Norian in favour of Juvavian. Very soon, in 1893, Bittner lodged his first protest. In a Reichsanstalt publication entitled "Was ist norisch ?" Bittner maintained that Mojsisovics was making improper use of the name.[132] To replace Mojsisovics' Archelaus Norian Bittner proposed the name Ladinian. Also published in 1893 was the largest part of Monograph 6 of the Reichsanstalt, the one Mojsisovics had started in 1873 as "Das Gebirge um Hallstatt", but now being called "Die Cephalopoden der Hallstätter Kalke". Mojsisovics had presumably decided he could not fulfil his original expectation, hence the change in the title. His health was begining to fail. In this monograph Mojsisovics gave a table with a Triassic chronology for the Mediterranean Province. Mediterranean province now embraced both the northern and southern Alps. In ascending order Mojsisovics had Norian, Carnian and Juvavian Stages. The Norian, of course, was Archelaus Norian. This was further provocation to Bittner, and in 1895 the Reichsanstalt published his review (145 printed pages!), on the recent Trias literature.[133] Most of this was devoted to analysing and criticizing Mojsisovics' different schemes. One wonders what the Reichsanstalt editor thought about this developing strife being aired in the official publication. For the remainder of this decade of controversy the officials were relieved of this problem. Mojsisovics published mostly in the proceedings of the Imperial Academy of Science, of which, by now, he was a full member. Starting in 1895

Bittner published most of his criticisms privately.[134] Eventually Mojsisovics was also in the private publishing business. In 1895, Mojsisovics, with Waagen and Diener, proposed a chronology for the whole Trias (Table III, page 60) taking into account data from Asia as well as the Alps.[127] Waagen and Diener's work is to the top of the Anisian; Mojsisovics did the remainder. Archelaus Norian and Juvavian appear in Mojsisovics' contribution, as in the 1893 Table.[126] Mojsisovics' schemes for 1893 and 1895 are much the same, except in the later one he introduced names for his substages, Upper Juvavian became Sevatian; Middle Juvavian, Alaunian, Lower Juvavian, Lacian and so forth. In 1896 Bittner made an aggressive criticism in his second private publication.[135] He seems to have had his cake and eaten it, because he reviewed and summarized his own private publication in the official Reichsanstalt report ![136] In 1896 Mojsisovics published a paper[137] which provoked a third privately printed criticism from Bittner.[138]

Matters really came to a head in 1898. So much happened that one must track the events month by month and day by day. Early in the year Bittner printed a pamphlet entitled "Herr E. von Mojsisovics and Ethics".[139] This was a passionate personal attack on Mojsisovics implying that his position as a member of the Imperial Academy of Science, and as Vice-Director of the Reichsanstalt was not justified. In February two letters were sent to Mojsisovics. They were printed, and obviously intended for circulation. The first, dated the 24th, bears the name of Eduard Suess and 34 other full members of the Imperial Academy of Science.[140] This letter makes no specific reference to Bittner and the controversy but records the respect that Mojsisovics' fellow academicians have for his scientific integrity. There can be no doubt that they wrote this letter to reassure Mojsisovics of their confidence in the face of Bittner's recent attack. The second letter, on February 26, was also supportive and was from Wilhelm Waagen and nine other corresponding members of the Academy.[141] This letter specifically refers to Bittner and his latest pamphlet. On March 6 Bittner printed a reply to the 35 academicians,[142] despite the fact that their letter was not addressed to him ! He levelled assorted charges of malpractice against Mojsisovics, and stated that if the academicians did not provide an answer, it was tantamount to their admitting Mojsisovics' guilt. Bittner's reply of March 6, although bearing that date, for some reason was not distributed until about a year later. In any event the 35 do not seem to have answered Bittner's accusations. Some time in March Mojsisovics printed a document entitled "A defence against Dr. Alexander Bittner".[143] In this Mojsisovics admits that he has made mistakes in his efforts towards deciphering the difficult geology of the Alpine Trias but takes exception to the aggressive and personal nature of Bittner's attacks. This did not draw an immediate response from Bittner.

Bittner did not really need to reply at this stage because at the end of March the bombshell exploded ! This was a printed document entitled "Zur Ordnung der Trias-Nomenclatur".[144] Translated freely this means: "On the business of Trias Nomenclature". This document is highly critical of Mojsisovics, and supportive of Bittner. It was signed by no less than 48 Austrian geologists. Bittner himself was not a signatory. The list includes several Trias specialists. One, von Hauer, we have already met. Others: Gustav von Arthaber, Ernst Kittl, Albrecht von Krafft, Franz Toula, Wilhelm Waagen, we will meet on later pages. Guido Stache (1833-1921), who was Director of the Reichsanstalt at the time signed the document, as did Emil Tietze (1845-1931), who subsequently achieved the same office. Most of the remainder were jackals, not Trias specialists. The scientific community was clearly polarized, with most of the Reichsanstalt staff backing Bittner and the Academy people siding with the Reichsanstalt Vice-Director – Mojsisovics. One who signed was August Böhm Edlen von Bömersheim (1858-1930). This fellow was not a Triassic specialist, but as we will soon see, was very well read on Triassic matters. When he died he was described as being particularly capable of wielding a sharp pen in criticism.[145] He seems to have been a real trouble-maker because he not only signed "Zur Ordnung. . ." but was also the publisher !

Distribution of "Zur Ordnung. . ." was followed, on March 28, by a short printed letter from Rudolph Hoernes (1850-1912) to Böhm.[146] Rudolph was the son of Moritz and thus a member of the Suess family circle (Chapter IV). At this time Rudolph was a professor at Graz. Previously he had worked with Mojsisovics at the Reichsanstalt. Hoernes was not impressed by the number of signatories for "Zur Ordnung. . ." remarking that in scientific questions the majority does not necessarily overule authority, the authority in this instance being Mojsisovics. April 12 brought a letter to Mojsisovics, signed by Carl Diener, C.M.Paul, Rudolph Hoernes, Ed. Reyer and Eduard Suess.[147] This letter also makes reference to "Zur Ordnung..". Mojsisovics was addressed as "Hochgeehrter Herr !" (Highly Esteemed Sir !). Like the February letters it was not intended for his eyes alone. It was printed, and obviously for circulation. Those who signed it had not signed "Zur Ordnung ..". They were Mojsisovics' staunch allies, and they beseeched him to cool the situation by dropping the name Norian from the chronological scheme. On April 14, Mojsisovics, provided a reply, also printed, to Suess, addressed as "Hochverehrter Freund !" (Highly admired Friend !). In his reply, Mojsisovics agreed to drop the Norian from the hierarchy. But if he thought that this was the end of the matter he was to be disappointed. Before the end of 1898 Böhm the trouble-maker had written and published a 31-page booklet entitled "Justice and Truth in the Nomenclature of the Upper Alpine Trias".[148] He would not accept the Mojsisovics compromise. "Between right and wrong there can be no pact"; so said Böhm. Mojsisovics, he maintained, was wrong. Metternichi Norian must stay; Juvavian must go.

In 1898 the names of no less than 98 scientists were associated with the controversy. Until March Mojsisovics had 49 allies. With the distribution of "Zur Ordnung. . .", Waagen, who had been supportive on February 26, was revealed as a renegade – a supporter of Bittner. So from March the camps were even: 48 on each side. Adding the protagonists we arrive at the total of 98.

The affair dragged on through 1899 and 1900. In April 1899 Bittner resumed his attacks.[149] In the same year Professor Zittel published his famous history of geology and paleontology, the book that was soon afterwards translated into English by Maria Ogilvie-Gordon (Chapter XII). In November 1899 Bittner distributed a pamphlet in which he takes Zittel's account of the polemics as support for his own case.[150] Bittner's Christmas mail included a copy of a letter from Zittel to Suess, dated December 24. This had been printed and distributed by Suess.[151] In this letter Zittel disassociates himself from Bittner's interpretation and deplores the personal nature of Bittner's attack. This gave Bittner something to do over the Christmas Season at the turn of the Century. On January 8, 1900 he had written a reply to Zittel, amounting to 10 printed pages. With this, Bittner also printed a supportive letter from Böhm the trouble-maker.[152] He followed up with another 4 pages on March 15.[153] This one was partly provoked by a review of Zittel's book in a Vienna newspaper, the "Wiener Zeitung" of March 9. The reviewer had noted some nice things that Zittel had written about Mojsisovics. This was too much for Bittner, who wrote that this was an unwarranted advertisement for Mojsisovics' achievements.

On October 20 1900 Mojsisovics was deprived of his position as Vice-Director of the Reichsanstalt. Poor health was given as the grounds for his retirement but Bittner thought it was for a different reason. Apparently there had been a disciplinary investigation into the matter between January and September. Bittner was called to defend his position. He seems to have been vindicated and expressed the opinion that his victory was the real reason for Mojsisovics' retirement.[153] Mojsisovics was replaced as Vice-Director by Tietze, one of the "Zur Ordnung. . ." signatories. Mojsisovics had been given the honorary title "Hofrath" on June 25 but his name is conspicuously absent from the 1900 Reichsanstalt staff list. This included everybody: at the top, the Director, at the bottom the porter and the man responsible for the heating, but no Mojsisovics. Bittner was still listed as a Chief Geologist. His spite and vitriol seem to have been effective. One might suppose that he was satisfied at last.

But Bittner was still not prepared to leave Mojsisovics in peace. In 1902 Mojsisovics, although retired, published a short paper in the second issue of the Reichsanstalt report – the Verhandlungen.[154] It described a Triassic fossiliferous occurrence and Mojsisovics used the term "lacisch". Soon afterwards, in the fourth issue, dated March 4, Bittner was in action again, asking "Was ist lacisch ?".[155] Was he preparing for another onslaught ? We will never know, because before the month had ended he was dead. Bittner concluded that "lacisch" was Lower Norian; to Mojsisovics it was Lower Juvavian. Regarding the real position of the lacisch (or Lac) within the Upper Triassic it is now known that both Mojsisovics and Bittner were wrong, whether one chooses to call it Norian or Juvavian (Chapter II).

In retrospect the whole affair seems both ridiculous and tragic. It is astonishing that it was considered so serious at the time, with the power to divide the Vienna scientific community in the way that it did. It is also astonishing that Bittner's vindictive behaviour was condoned and indeed encouraged by so many of his Reichsanstalt colleagues. I can see absolutely no justification for Bittner's attitude and actions. He was quite unnecessarily aggressive and offensive in his attacks on Mojsisovics. His polemics contributed nothing to the science. Mojsisovics worked on the Trias for many years. He made mistakes and managed to correct some of them. But all along he was working towards deciphering the story. He seldom if ever wrote an unkind word. For several years he paid little attention to Bittner's attacks. This infuriated Bittner and made him increasingly offensive. Perhaps Bittner did not make many mistakes, but this is because his own contributions to Triassic chronology are trivial compared with Mojsisovics'. Bittner named the Ladinian Stage, but it was based on stratigraphy elucidated by his predecessors.

Georg Rosenberg (1897-1969),[156] who knew the Alpine Trias well, has provided a just and generous appreciation of Mojsisovics' work, viewed in the perspective of half a century since he died.[157]

Only once did Mojsisovics publish his compromise solution. This was in the final, supplement volume of the Volume 6 Monograph.[158] In this, the upward succession of stages was Ladinian, Carnian, Juvavian. He had met Bittner half way, by using Ladinian. Bittner never saw this in print; he died on March 31, 1902. The Monograph was published on July 1. But Mojsisovics' 1902 scheme never caught on. In 1903 the International Geological Congress met in Vienna. Ernst Kittl (1854-1913) of Vienna, who had wide experience on the Alpine-Mediterranean Trias, wrote the guide for the Salzkammergut field trip. In this, Metternichi Norian is reinstated. In 1905, Arthaber did the same in the "Lethaea", a book that will be mentioned in Chapter XII. Mojsisovics died in 1907. Five years later, his old ally, Carl Diener, had also turned against him.[159] So in the end, August Böhm Edlen von Bömersheim, whose name is hardly known to workers on Trias, exercised more influence than the great Eduard Suess!

IX

Refinement in Asia: Salt Range and Himalayas

In Chapter VI we saw that Trias discoveries had been made as early as 1825 in the Himalayas and 1851 in the Salt Range. There were thus ample grounds to suspect that Asia might be the source of critical data for Triassic time scale. In the last century these areas were part of the British Empire. The Geological Survey of India was founded in 1851 under this British administration. Staff for the Survey came not only from Britain, but also from the Vienna school of geologists and paleontologists. Between 1869 and 1902 enormous contributions to the marine Trias were made under the auspices of the India Survey, mostly by Vienna alumni. So in Asia, as in Europe, Vienna dominated the field. Most of the results appeared in great quarto volumes, Palaeontologica Indica, published by the Survey. The volumes of this series, along with the monographs of the Reichsanstalt in Vienna, form the classics for the marine Trias.

The Salt Range,in what is now Pakistan, was first examined in detail by Arthur Beavor Wynne (1835-1906) of the Survey, between 1869 and 1872. Wynne was not a paleontologist. Because fossils were very abundant he was latterly joined in the field work by Wilhelm Heinrich Waagen, who was born in Munich (1841) and died in Vienna (1900).[160] Between 1870 and 1875 Waagen was employed by the India Survey as a paleontologist. Waagen's work led to two massive contributions to Palaeontologia Indica; one on the Permian (Productus Limestone Fossils), the other on the Trias (Fossils of the Ceratite Formation, 1895). This was exciting stuff. The Ceratite Formation gave an indication of what the Lower Triassic ammonoid faunas looked like. It was a part of time for which there was very poor representation in Europe. Waagen's work was careful and detailed. In a later chapter we will see that there have been a few subsequent discoveries, but not many. His work was a land-mark.

Coming now to the Himalayas we encounter a succession of workers who made notable achievements.

First was Ferdinand Stoliczka.[161] Stoliczka was born in Moravia in 1838. As a young man he studied in Vienna under Suess. In 1862 he went to India, to work for the Geological Survey. Stoliczka was a good paleontologist and all-round naturalist, much liked by his friends and colleagues. In 1864 he went to Spiti with his friend and colleague, F.R.Mallet. Mallet and William Theobold, another Geological Survey colleague, had been there in 1861 and brought back collections that indicated the potential. Stolickza did the pioneer work on the very important stratigraphic section on the northern slope of the Himalayas. He worked through both the Paleozoic and Mesozoic. He made a mistake by recognizing only the Upper Triassic, and missing the Lower and Middle. But he laid good foundations for his successors. Stoliczka also did important work on the Jurassic and Cretacous, at Spiti and in other parts of India. In 1873 he had to make a difficult decision. There was to be an

HIMALAYAN TRIAS

Profile of the upper Girthi Valley, in the Niti area, sketched by C.L. Griesbach in the 1880s. A concordant section from the Ordovician to the Jurassic is exposed. The Permian-Triassic boundary is between 9 and 10. 10, Lower Trias; 11, Middle Trias; 12-15, higher beds of the Trias; 16, Jurassic. The total thickness of the Trias is about 1000 metres. From Memoirs of the Geological Survey of India, v. 23, 1891.

International Exhibition in Vienna. Stoliczka had the option of going there, with a collection of fossils and minerals, to show the scientific world what had been achieved by the Geological Survey of India. But there was to be another, conflicting, invitation.

The British rulers of India were planning a second diplomatic mission to Yarkand and Kashgar. Thomas Douglas Forsyth (1827-1886)[162] had already been there in 1870, on the first mission and he was appointed to lead the second. Forsyth was born and died in England, where he was educated at Rugby School. Most of his life was spent in India; until the Mutiny (1857) with the East India Company, later with the British government, finally as a Director of railway companies.

Yarkand and Kashgar, although dominantly Moslem, were supposedly within the Manchu Empire. Their ruler, in fact, was Yakub Beg (1820-1877), who was also known as the Atalyk Ghazi.[163] Yakub Beg was in power, with Kashgar as his base, from 1864 until he was assassinated in 1877. His desire for independence was encouraged by the Turkish Sultan, who in 1873 conferred upon him the title Amir. He did not like his Russian and Chinese neighbours but was interested in good relations with the British. This is shown by the friendly reception he gave to George Hayward in 1869 when he encouraged future visitors.[164] However on the occasion of Forsyth's first visit, in 1870, he was engaged in some way and the two did not meet. Hayward records that the Amir had, incidentally, avenged the murder of Adolf Schlagintweit by cutting the throat of the perpetrator, Wullee Khan, in 1867. Hayward, sad to say, was himself murdered at Yasin, in the Hindu Kush, soon after his encounter with Yakub Beg.[165]

The object of the Yarkand Mission was a commercial treaty. Maintaining good relations with Yakub Beg was also desirable in order that he might remain a buffer between the Russian and British Empires. The route from India to Yarkand was not easy. The Himalayan and Karakoram Ranges had to be crossed. It became an expedition and it was decided to take a naturalist. Stoliczka was invited. He decided on the expedition instead of Vienna. It was to prove an unfortunate decision. The party left Calcutta in May 1873. Stoliczka formed a detached party which took a route across the Sanju Pass (4,950 m). Here Stoliczka showed symptoms of serious illness. His companions thought he was suffering from an attack of spinal meningitis. He carried on, arriving at Yarkand on November 8, Kashgar on December 4. Here he was able to rest until January 10, 1874, when Yakub Beg received the mission. The treaty was settled. Following the mission, around 1877, Andrew Dalgleish, a British trader, settled in Yarkand. He met the fate of Schlagintweit and Hayward, by being murdered by an Afghan on the Karakoram Pass in 1887.[166] Not many Europeans had visited this corner of Asia without being murdered. Being infidels in lands where Mohammedans held sway they were particularly vulnerable. Colonel Stoddart and Captain Connolly established the pattern at the hands of the Amir of Bokhara, in 1842. The toll, to 1877, was thus at least five – Stoddart, Connolly, Schlagintweit, Hayward and Dalgleish.

For the success of the mission Forsyth was honoured with a knighthood – Knight Commander of the Order of the Star of India.

On the return journey the party took the Karakoram Pass, more than 5,400 m high. On June 16 they were at the summit. Stoliczka again became ill. On the 17th he made his last diary entry; on the 19th he died. Some eleven marches later his companions buried him under a willow tree at Leh, Kashmir.[167]

Stoliczka was busily doing geological work until the day before he died.[168] He sent back reports to the Geological Survey describing his observations on the outward journey.[169] His good friend, W.T.Blanford, wrote an account of the work done while returning.[170] Collections of *Monotis* and *Heterastridium* proved the occurrence of Norian on the northern approach to the Karakoram Pass.[171] *Heterastridium* is a curious Triassic fossil, widely distributed in Norian deposits. Sphaerical, most look like golf balls but some are like marbles or baseballs. They may have been some sort of floating coral-like organism. They often occur abundantly. This was evidently so where Stoliczka found them, because they were already known to the local people as Karakoram stones.[172]

After Stoliczka, the next to appear on the scene was Carl Ludloph Griesbach.[173] Born in Vienna (1847), for a time he worked with the Reichsanstalt. In 1878 he joined the Geological Survey of India, and between 1894 and 1903 was its Director. His achievements were acknowledged when he was made a Companion of the Order of the Indian Empire (C.I.E.). Griesbach spent several years doing field work in the Himalayas, all the way from the Nepal frontier, west to Spiti. With his work the importance of the Himalayan section really became apparent. Griesbach discovered *Otoceras*. Although more a geologist than a paleontologist Griesbach described this important fossil himself. He was the first to provide ammonoid data that bear on the question of defining the Permian-Triassic boundary, and gave full justification for the conclusion that the *Otoceras woodwardi* Zone should be taken to mark the base of the Trias. He was an accomplished artist and lithographer, and his description of the ammonoid is accompanied by his own excellent lithographs. Before he went to India his skills in lithography were recognized and appreciated, to the extent that his talents were employed illustrating fossils in paleontological papers written by other scientists.[174] When Griesbach published his classic – Geology of the Central Himalayas – in 1891, it was clear that the Himalayas revealed a succession for the whole marine Trias, with incomparably better sequences of ammonoid beds than those preserved in the Alps. This report includes many of Griesbach's own excellent field sketches showing the topography and geological structure. Griesbach died in Graz, in 1907.

The Vienna scientists, both those in residence and those expatriate, found Griesbach's discoveries very exciting. A joint expedition of the Imperial Academy and the Geological Survey took place in 1892.[175] Griesbach was on it, and also Carl Diener, who was soon to become a prominent figure in Trias studies. Diener (1862-1928)[176] was born and died in Vienna. In his early University days he was more of a geographer than a geologist. Later he became a professor of geology and paleontology in Vienna. The 1892 expedition made large collections which were later described by Mojsisovics and Diener himself. In Tibet they discovered the "exotic blocks" – masses of red limestone, virtually indistinguishable from the Hallstatt Limestone of Salzkammergut. Today these limestone blocks are only a few kilometres from the bedded sequence that extends in a belt from Spiti to Nepal and Sikkim. They are in the strip now known as the Indus suture, associated with rocks that were probably part of the Tethys oceanic floor in the Trias. So in Triassic time the sediments now forming the exotic blocks were probably hundreds of kilometres distant from the sediments preserved in the main Himalayan ranges.

In 1936 the area with the exotic blocks was visited by a remarkable Swiss expedition.[177] It consisted of only two people, Arnold Heim (1882-1965) and Augusto Gansser. Although remarkably small it was also remarkably successful. They started with a third member, Werner Weckert, but he had to drop out owing to an attack of appendicitis.

When Heim and Gansser reached India they were assisted in their planning by another Himalayan geologist, John Bicknell Auden, of the Indian Survey. His brother was the poet, Wystan; my mother is their first cousin. Auden had joined the Geological Survey in 1926, soon after graduation from Cambridge University. Between 1932 and 1934 he made several traverses of the Himalaya. In 1934 he became one of the first geologists to visit the Kingdom of Nepal. Normally it was very difficult for Europeans to gain access to Nepal in those days, which is why all the early Mount Everest expeditions had to approach the mountain from Tibet. Auden was officially invited to Nepal to investigate the geological aspects of the Bihar-Nepal earthquake of January 15, 1934. This was very severe, 8.4 on the Richter

Scale. 10,700 people died. Auden's work in Nepal did not take him to the northern part, where the Triassic is well developed. This would come to light from the work of Heim and Gansser in 1936, and from work done by Austrian and French expeditions starting in the 1960s. By that time Nepal was relatively accessible. Important discoveries were made by Gerhard Fuchs and René Mouterde. Their pioneering work has been followed by the detailed investigations of L. Krystyn (Chapter XVI). Although Auden did not reach the Trias belt in Nepal, later in 1934 he was to discover it on the Sikkim-Tibet frontier. This was on a journey undertaken when he was on leave. He crossed the Himalaya in Sikkim, which at that time was the Kingdom immediately east of Nepal. Nowadays it is a province of India. In northern Sikkim, at the Tibet frontier, Auden discovered rocks with Middle Triassic ammonoids at Tso-Lhamo Lake.[178] This was important for indicating the eastward extension of the Trias belt which had been made known, west of Nepal, by the work of Griesbach and others who had been Auden's predecessors on the India Survey.

Returning to Heim and Gansser. They were supposed to confine their activities to the parts of the Himalayas that lay within British India. Unlike Auden they had no invitation to Nepal. But when they reached the frontiers with Nepal and Tibet they could not resist the temptation to trespass into these forbidden territories to see the exotic blocks and other interesting things. In making these forays they were arrested twice, but managed to get off and continue their work. Gansser was particularly intrepid. On one trip to Tibet he was disguised as a pilgrim outbound and as a shepherd on the return. He adopted the disguise of a shepherd because the Tibetans used sheep and goats as pack animals for taking salt to India and for bringing back sugar, rice and flour. Gansser in his role as a shepherd smuggled out his specimens in salt bags. From the exotic blocks (Kiogars) they brought back Carnian ammonoids from red Hallstatt-type limestone. They also collected ammonoids at Tinkar Kipu, from dark coloured limestones of the main Triassic belt. This locality is where the frontiers of India, Nepal and Tibet meet, and the acquisition of the fossils necessitated an illegal foray into western Nepal. The fossils were eventually brought back to Zurich, where they were described by their colleague Alphonse Jeannet (1883-1962).

The exciting early period of Himalayan Trias studies had come to an end with the work of Albrecht Krafft von Dellmensingen.[179] Born in Franconia (1871), he was a student under Zittel at Munich, later at Vienna, where he worked with Suess and Waagen. He joined the Geological Survey of India late in 1898. In 1899 he accompanied Hayden to Spiti. Henry Hubert Hayden (1869-1923) (eventually Sir Henry), who wrote the classical account on Spiti geology, became Director of the Indian Survey in 1900. In 1900 von Krafft spent a very long season in the Himalayas, accompanied by his courageous young wife. In these two years Krafft restudied virtually all the localities known to Griesbach, Hayden and Diener, and gave close attention to the exotic blocks. Many new finds of fossiliferous beds were made. He was scrupulously careful in his collecting. Fossils from beds separated by a centimetre or two were kept apart. Sad to say, on September 22, 1901 he died suddenly in Calcutta, of heart failure. Presumably, as with Stoliczka, it was strenuous activity at high altitude that did him in. Before he died he wrote a short account for the Geological Survey describing the full succession. It would fall to Diener to actually describe most of Krafft's fossils. As a result of Krafft's untimely death, Diener tends to be regarded as the doyen of the Himalayan Trias despite the fact that he did field work only on the 1892 expedition. He only knew the surroundings of the Niti area at first hand. He was never at Spiti (to the west), nor at Byans (east). But Diener was fair. In 1902 he wrote a generous appreciation of Krafft's work.[180] Diener recognized that with Krafft's contributions, the Spiti section had become the best in the world for displaying Trias ammonoid faunas in sequence. It is probably true today. Northeastern British Columbia is its only rival.

NORTH AMERICA

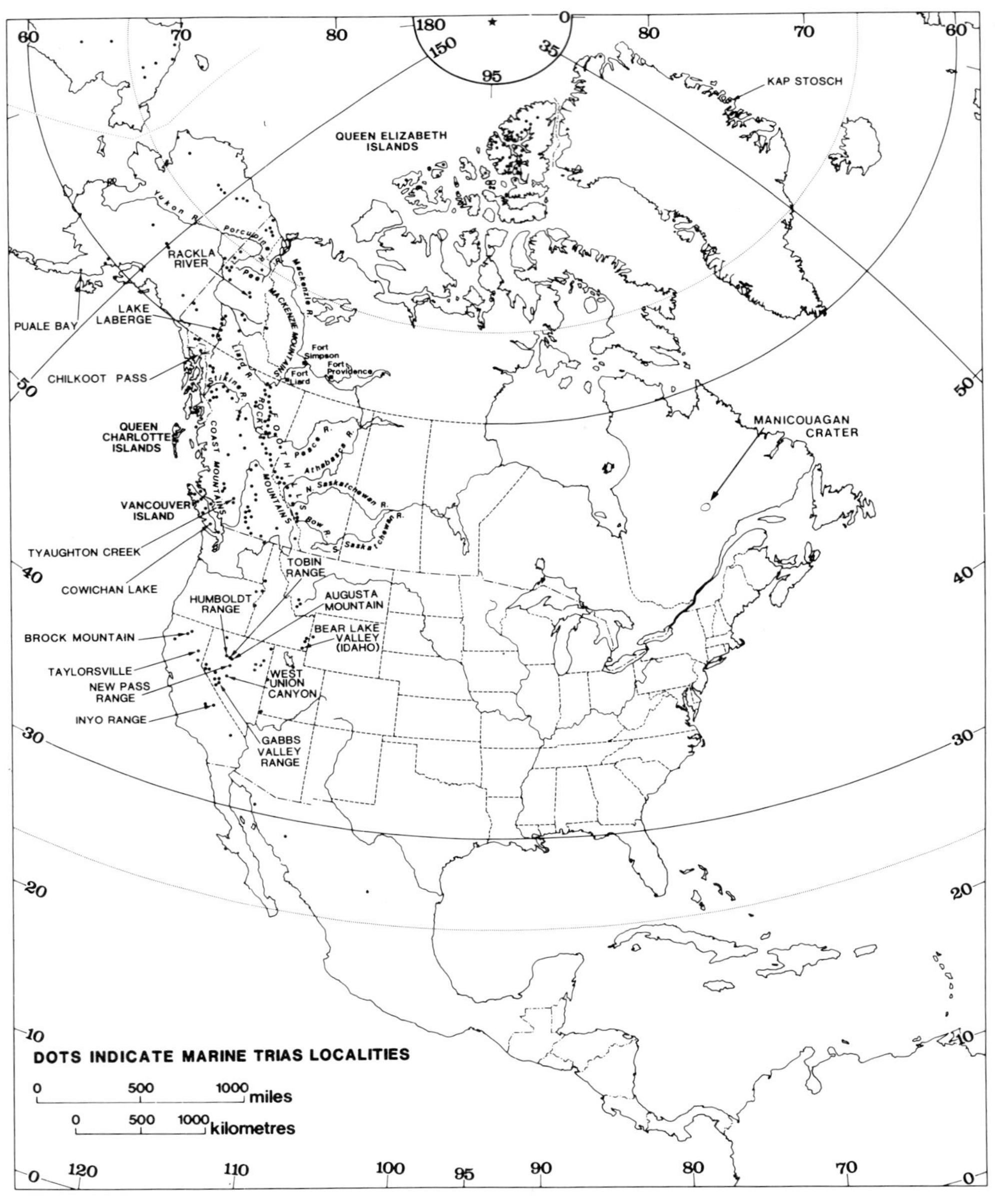

X

Progress in America: Hyatt and Smith

In Chapter VI we saw that by 1890 it was known that marine Trias was widely distributed in the American Cordillera, from California to Alaska. Fossils of Lower, Middle and Upper Triassic age had been found. But because they came from widely scattered localities and sequential relations were unknown, the American data did not contribute much to the time scale. In the decade from 1894 great progress was made, largely through the work of the Stanford professor James Perrin Smith (1864-1931), who was known as "JP".[181]

The best known publication for this period is "The Triassic Cephalopod Genera of America" authored by Alpheus Hyatt and JP, and published in 1905. Hyatt died in 1902, so it was posthumous as far as he was concerned.

Hyatt was born in 1838, of an old Maryland family.[182] As a young man he was for a while a student at Yale; later a student under Louis Agassiz (1807-1873) at Harvard. During the Civil War he was a Captain in the Union Army. He then became a dominant figure in the study of fossil cephalopods, of all ages and all kinds. His work was not only descriptive but also philosophical and theoretical. Because cephalopods preserve their development (ontogeny) in their shells, and because the ontogeny of all animals casts light on their evolution, Hyatt had a great time developing theories about cephalopod evolution based on their ontogeny. Some of his ideas may have been sound but now there is general agreement that he got carried away by drawing conclusions from ontogeny alone, and not paying attention to the facts regarding their succession in time. His base of operations was Boston, where he was a professor at the Massachusetts Institute of Technology and at Boston University. Later he was at the Museum of Comparative Zoology on the Harvard Campus. Besides working on cephalopods in Boston he also did some field work on the Trias of the West. In 1892 he wrote on the Taylorsville area, where Clarence King had found *Monotis* (Chapter VI). Hyatt mistakenly thought that beds with *Tropites* were higher stratigraphically than those with *Monotis*. He was having trouble with the Carnian-Norian relationship, the very same problem that confronted Mojsisovics in Austria (Chapter VII). Hyatt read his paper in 1891, the year before Mojsisovics produced his revision of the original Carnian-Norian relationship. Hyatt seems to have used Mojsisovics' original interpretation in formulating his own. An incorrect interpretation in Austria was being used to support an incorrect interpretation in California. About 50 years later F.H.McLearn would face a similar dilemma in British Columbia (Chapter XIV).

In the Hyatt and Smith monograph JP writes that it was inspired by Hyatt and done under his general supervision. But clearly JP did most of the work. The title "Hyatt and Smith" does less than justice to JP's own contribution. It may reflect JP's generous and amiable temperament which evidently endeared him to both colleagues and students.

One of JP's earliest students was Herbert Clark Hoover (1874-1964) who became the 31st President of the United States (1929-1933). Hoover graduated from Stanford in 1895

USA : HYATT AND SMITH

Stanford University Expedition to Trias of Northern California, 1893. "JP" is front right.

and for a while practiced as a mining engineer. He had scholarly interests, and with his wife (Lou Henry Hoover, 1875-1944) translated Agricola's "De Re Metallica".[183] He also engaged in a number of humanitarian international projects. It was his misfortune to preside over the United States at the onset of the Great Depression, for which he can hardly be held responsible. Hoover, of course, was a Republican; JP, a Democrat. When the time came for JP to cast his vote in 1928, he could not face voting Republican, but he was equally reluctant to vote against his old friend. He resolved the problem by abstaining.[184]

JP's role during this period of Triassic discoveries is apparent from his "The Comparative Stratigraphy of the Marine Trias of Western America", published in 1904. JP was born in South Carolina. His schooling and undergraduate studies were in the East. In 1890 he went to Germany for geological studies, eventually gaining a Ph. D. at Göttingen. His thesis was on the Jurassic stratigraphy and paleontology of an area near Echte, south of the Harz Mountains.[185] Muschelkalk was also exposed within the area he studied. The fact that he collected specimens of *Ceratites* from the Muschelkalk is testified by the existence of specimens that he brought back to the Stanford University collection. He went to Munich, where he met Zittel, but evidently did not go to Vienna to meet Mojsisovics. So he was not on hand to witness the start of the polemics with Bittner. On his return to the United States JP took up Trias ammonoids and stratigraphy in a big way. He also worked extensively on Carboniferous faunas. Between 1894 and 1904 JP did a lot of field work and collecting. For the Lower Triassic he worked around Bear Lake Valley in Idaho and in the Inyo Range, California; for the Middle in the Humboldt Range, Nevada and for the Upper around Brock Mountain, in northern California. In this way he acquired data for his 1904 paper.

He was certainly an active field worker during this period. Sometimes he went alone, at other times with his family. I had it from Frank McLearn that Charles Schuchert (1858-1942) did not approve of JP's family jaunts. Field work was serious business; families a distraction. Schuchert was a professor at Yale. McLearn heard of this disapproval when he was a Yale research student, under Schuchert's supervision, before the 1914 War. Schuchert's career was remarkable in that he became a Yale Professor without ever having attended University. He was trained as an apprentice to James Hall (1811-1898), the famous Paleozoic paleontologist whose centre of activity was New York State. Schuchert never had the opportunity to go on family jaunts. Preston Cloud told me that Schuchert's sweetheart, from teen-age years, married Edward Oscar Ulrich (1857-1944), another prominent figure in American paleontology. Schuchert remained a hard-working bachelor. So perhaps his disapproval of JP's family outings reflected regrets of what he had missed and might be missing, by not having a family of his own.

JP certainly collected lots of fossils himself. But he was not above acquiring them by other means. The rocks in Nevada that produce ammonoids are also the source of reptile bones, mostly of ichthyosaurs. In the early years of the 20th century these bones were being collected and studied by Professor John C. Merriam (1869-1945) of the University of California. Annie Montague Alexander (1867-1950) was a member of one of these expeditions, to the Humboldt Range, and she has left an account of an encounter with JP.[186] The California University Expedition included a Mr. Boynton. While hunting for bones, Mr. Boynton found an ammonoid "as big and as round and as flat as a plate". Mr. Boynton was delighted with his find, but also covetous. A visit from JP was expected. "You must show the specimen to the Professor", Boynton was told by his companions. Boynton had mixed feelings. His pride urged him to show it; but he was afraid he would lose it to the Professor, and he wanted to keep it. When JP arrived he decided to show it off. JP could not contain his excitement. "A *Gymnites*, a *Gymnites*" he exclaimed, "I never expected to see one in the flesh". But Boynton would not part with it. At night, for safety, he kept it under his pillow. Eventually Boynton and JP were together in a game of whist. Boynton evidently ran low on cash and had to stake the *Gymnites*. He lost. JP took it back to Stanford, wrapped, first in his night shirt, then a pair of pantaloons, and finally in his sleeping bag. It had been discovered too late for the Hyatt and Smith book but was eventually described in 1914. Described, one might say, both lovingly and unfairly ! Lovingly in that the illustration is more a restoration than a picture of what the ammonoid actually looks like; unfairly in that the name commemorates Annie, not Mr. Boynton. Smith named it *Gymnites alexandrae*; Mr. Boynton, it seems, was forgotten.

JP had started publishing about the American Trias in the 1890s. By this time he was in touch with Mojsisovics, exchanging specimens and data. Hyatt actually visited Mojsisovics in 1897. JP wrote his first Triassic papers in 1894 and 1896.[187] They were pretty muddled. He was trying to make use of the Mojsisovics-Waagen-Diener Trias chronology which was hot off the press (1895). Norian was Archelaus Norian and the term Juvavian was being used. The actual American succession was not clearly expressed. The confusion introduced by Hyatt, with *Tropites* allegedly above *Monotis*, was there. He more than redeemed himself with the 1904, Comparative Stratigraphy Paper. Here the details of the North American succession are clear. *Monotis* is correctly located above *Tropites*. Ladinian is used in place of Archelaus Norian and the original Norian restored. Curiously JP makes absolutely no reference to the battle between Mojsisovics and Bittner. Perhaps he regarded it as a tragic confrontation, damaging, quite unnecessarily, Mojsisovics' reputation. JP evidently was a kindly, good-natured man, who kept his science in the right perspective. Some of his Vienna counterparts were of a different temperament.

One can hardly speak too highly of JP's 1904 contribution. It was an able and enlightening review. The Table in the 1905 monograph also summarizes the succession, with one addition, the *Columbites* fauna. Some of JP's correlations have now been changed and many new levels have been discovered. But his sequences were right. He might be faulted for one thing. For the Middle Triassic of Nevada, where he had enormous collections from the Humboldt Range, fossils from what are now known to represent many successive different faunas were lumped as one Anisian Zone – the *Daonella dubia* Zone, but even here he was aware of the possibility that some Ladinian beds might be represented. Later work would show that this zone covers the whole of the Anisian and part of the Ladinian.

After 1905 JP produced three more books on the Trias. Middle Triassic appeared in 1914; Upper in 1926; Lower (published posthumously) in 1932. They add relatively little. For his last quarter century JP evidently devoted more time to his students and perhaps also to whist. Shortly after the 1914-1918 war, which brought hard times to Vienna, JP was in touch with Carl Diener. Diener sent him specimens and I believe that JP provided financial help in return. JP died of pneumonia at Palo Alto, California, on New Year's Day, 1931.

XI

Timor Bonanza: Fossils without Stratigraphy

Early in this century it was discovered that Triassic ammonoids occurred in the East Indies. The best localities were in western Timor, now a part of Indonesia, then a Dutch colony. Several expeditions went for collections. Some were organized by the Dutch government, others by Johannes Wanner (1878-1957).[188] Wanner, who was born in Bavaria, travelled extensively in the East Indies between 1902 and 1911. Later he became a Professor at Bonn. As far as the Trias is concerned his most important expedition was in 1911, when he was accompanied by Otto Welter and C.A.Haniel. The ammonoids they found were astonishingly well preserved and abundant. They are preserved in reddish limestones, some with black manganiferous coating, almost exactly like those of the exotic blocks in Tibet and the Hallstatt limestones of the Mediterranean area. Collections were also made by Dutch expeditions in 1910-1912. These were brought back to the Museums in Bonn and Delft. The British Museum of Natural History in London also acquired a substantial collection from various sources. Timor thus proved an absolute bonanza for Triassic ammonoids. More properly it should be described as a large number of bonanzas, because all the ammonoids are from blocks, some little bigger than pebbles, others as big as a house, all embedded in a clay deposit of Cenozoic age. Nowadays formations like this are called olistostromes. The ammonoids from these blocks were described in a series of monographs by Welter, Diener and Arthaber. Faunas of Lower, Middle and Upper Triassic age were recognized. The monographs are books with pictures of beautiful ammonoids. But from a scientific point of view they do not contribute very much. Most of the specimens were collected by geologists who put all the ammonoids from one block in the same sack. It was recognized by Welter, and by H.D.M.Burck (one of the Dutch geologists) that some of the blocks showed stratification, with different fossils in different layers. But the exact layer from which most of the fossils were obtained was generally not recorded, or at any rate it is not recorded in the published literature. So the early discoveries in Timor produced marvellous collections but added little to the Trias biochronology.

In those days it was not realised that Hallstatt-type limestones accumulated very slowly. It is now known that a metre or so of this kind of limestone may provide a record of two or three million years of sedimentation. A block one metre thick has now been described in which there are no less than nine successive layers with different ammonoid faunas.[189] In Chapter XVI we will see that the sequential data preserved in these blocks is now realizing its potential for refining the time scale.

It is a pity that Otto Welter did not write more about the blocks. Of those who described the ammonoids he and Wanner were the only ones who actually went there. Welter's description of a block that may have successive layers covering the interval at the Lower-Middle Trias boundary suggests that he was aware of their potential. Diener and Arthaber, who might have been expected to derive benefit from the block stratigraphy, were never there. After writing his monographs Welter gave up the Trias. Bernie Kummel told me that they met in South America, in the 1950's. They spoke of Triassic ammonoids over a drink. Welter, by then, had apparently lost interest; indeed Bernie quoted Welter as having remarked that he now regarded the years he had spent studying Triassic ammonoids as completely wasted time. Bernie also remarked that Welter was then devoting most of his time to consuming alcohol. How long he survived I do not know.

XII

1900-1928: Syntheses and Revelation

By the first decade of the 20th century a lot was known about the marine Trias of Europe, Asia (particularly India) and America (particularly the United States). Canada had not yet revealed its riches. But even so it was a good time for stock-taking and synthesis. Three people involved in such undertakings were Maria Ogilvie-Gordon, Fritz Frech and Carl Diener.

Maria Ogilvie-Gordon (1864-1939)[190] is best known as the translator of "History of Geology and Palaeontology to the end of the Nineteenth Century" by Karl Alfred Ritter von Zittel (1839-1904), the Munich professor, who was perhaps one of the most versatile and scholarly geologists and paleontologists of all time. In this translation, which was published in 1901, there is a comprehensive account of early work on the Trias which is thoroughly authoritative, not surprisingly, because both Zittel and Ogilvie-Gordon had worked extensively on the subject. Maria Ogilvie was a remarkable women with several notable "firsts" to her credit. She was born in Aberdeen in 1864. Early on she considered a career in music, but later turned to science, first at the Heriot Watt College in Edinburgh, later at University College London, where she gained a B.Sc. in 1890. She then went to Munich where she worked with Professor Zittel, doing systematic work on fossil corals. For this she earned a Ph.D., the first woman to do so at Munich. She then met, and became the friend of Ferdinand Freiherr von Richthofen and his wife. Richthofen had done important work towards interpreting the reefs in the Trias of the Dolomites some 30 years earlier (Chapter VII). With his encouragement Maria took up work in the Dolomites and in 1892 presented an important paper on the subject to the Geological Society of London. For this, in 1893, she became the first woman to receive a D.Sc. at the University of London. The London D.Sc., then and now, was awarded for proven published research, not necessarily directed by a member of the University. In 1895 she married Dr.John Gordon, an Aberdeen physician and they had two daughters and a son. She was greatly distressed when he died in 1919. She continued doing geological work in the Dolomites after her marriage and by 1914 had prepared a substantial report for publication. The manuscript was in Munich at the outbreak of war. She returned to Munich in 1920 to find that the manuscript had disappeared. Undaunted, she prepared another, which was published in 1927 as a monograph of the Austrian Bundesantalt. This publication is something of a paradox. When Maria did the field work the Dolomites were within south Tirol, under Austrian administration. By the time that her results were published, after the 1914-1918 war, south Tirol was a part of Italy. At first sight it is surprising that the Austrian government published the report, but the explanation seems to be that Maria herself paid for the printing.[191] Besides being a geologist and paleontologist, she was also a social activist, and became a Justice of the Peace, one of the first women to do so. In 1919 she formed the council for the Representation of Women in the League of Nations. She was torn between her social and geological interests. In 1936 she wrote to her friend Julius von Pia (1887-1943),[192] the Vienna Trias specialist, "I keep toiling hard at public work, meetings of all kinds, while knowing inwardly that it is geology in my heart that I wish to do."[193] For many years she felt that her geological work was not appreciated in England but was perhaps satisfied when she received the Lyell Medal of the Geological Society of London in 1932. In 1934 she was elected Honorary Life President of the National Women Citizens Association and in 1935 created a Dame of the Order of the British Empire (D.B.E.). She also received an honorary Ll.D from the University of Edinburgh. She died in England, June 24, 1939.

Fritz Frech (1861-1917),[194] who became a professor at Breslau, had broad geological interests which included the Trias. He died of malaria during the 1914-1918 war, in Turkey, while a geologist in the German Army. Frech's own contributions to Triassic geology and paleontology are not outstanding. He is mostly known for organizing and editing the Trias volume of "Lethaea geognostica".

Lethaea and her lover, Olenus, make an all too brief appearance in Book X of Ovid's "Metamorphoses". He describes them as "once fond lovers, now stones set on well-watered Ida, all because Lethaea was too confident in her beauty, while Olenus sought to take the guilt upon his own shoulders, and wished to be considered the culprit". Ovid seems to assume that his readers would be familiar with the story and regrettably says no more. According to another Roman poet, Lucius Valerius Flaccus, Olenus had a son, but I have not discovered any more about the family. Ovid's remarks suggest that some sort of a beauty contest may have been involved. Mount Ida, where Lethaea and Olenus were petrified, was the site of the most famous beauty contest of all time, when Paris was called upon to pick a winner from three gorgeous goddesses: Hera, Athene and Aphrodite. Paris chose Aphrodite; she got the golden apple, Paris was rewarded with Helen. Despite the fact that both events took place at Mount Ida, Jay Macpherson, an authority on these matters, assures me that they are unrelated.

Lethaea's petrifaction was commemorated by early 19th Century paleontologists when they adopted her name for books on fossils. Most famous of these was the Lethaea geognostica. This was the title chosen by Bronn. Elegant and concise, it is a good substitute for the full title – "Abbildung und Beschreibung der für die Gebirgs-Formationen bezeichnendsten Versteinerungen" – which is quite a mouthful! Bronn's first edition was published between 1834 and 1838. A later edition, with three volumes of text and an Atlas of 124 plates, came out in 1851-56. In this there are five plates devoted to the Trias. They give an inadequate impression of the Triassic fauna, the Alpine faunas being poorly represented. *Halobia* and *Monotis*, which Bronn had described himself (Chapter IV), are missing as are the St. Cassian and Hallstatt ammonoids, despite the fact that many had already been described in the 1830s and 1840s. Bronn did not appreciate the full impact of Hauer's masterly interpretation of the Alpine Trias (Chapter V).

Bronn died in 1862 but the Lethaea lived on. Carl Ferdinand Roemer (1818-1891) had been associated with Bronn in connection with the 1851-1856 edition. Roemer was a German who worked extensively in both Europe and North America. He's been called the father of Texas geology. Roemer carried on with the Lethaea after Bronn's death, publishing a new atlas for the whole Paleozoic in 1876. Frech became associated with Roemer, and took over after he died, with publication starting in 1897.

The Trias volume of the Lethaea is in four parts, published between 1903 and 1908. It is a massive summary, more than 600 pages, with numerous maps, sections, charts and illustrations of fossils. It remains a very useful summary of the work done to the turn of the century. Parts were written by Frech himself. Emil Phillipi (1871-1910)[195] did the continental Trias and Muschelkalk; Gustav von Arthaber (1864-1943),[196] the Mediterranean Trias; and Fritz Noetling (1857-1928 or 1929),[197] the Trias of Asia. Phillipi, mainly known for his monograph on the Germanic Ceratites worked in Jena and Berlin. Arthaber who became a professor in Vienna, did extensive work on Triassic faunas from Austria, Italy, Hungary, Albania and Turkey. Noetling had a bit of first hand knowledge of the Salt Range and Himalayas, having worked there for the Geological Survey of India around the turn of the Century.

The Lethaea, being a summary and synthesis, did not provide many new data. There are, however, some interpretations which deserve comment. The *Otoceras* beds of the Himalayas are excluded from the Trias. They are dealt with in a companion volume on the "Dyas" (= Permian). This has to do with the question of the Permian-Triassic boundary. I say question advisedly; often this is spoken of as the boundary problem. There is a problem concerning the significance of the boundary. But the definition of its position is simply a matter of convention and priority. Griesbach was the first to provide useful data towards defining the boundary in such a way that it would be widely recognizable. Let us continue to abide by his decision (Chapter IX). The view of Frech and Noetling that the *Otoceras* beds be regarded as Permian did not gain acceptance in the early years of this century. For about 50 years it was forgotten and ignored. Only in the last few years has the issue been exhumed.[198] It is better left buried.

Arthaber's account of the Mediterranean Trias is significant in that Mojsisovics' zonal scheme for the Upper Triassic, particularly the Norian, is given a very low profile. Mojsisovics' "zones" are tabulated as "lenses". The sequential arrangement is questioned, as it had been in the 1903 International Congress Guide by Kittl. Despite these doubts, expressed so long ago, Mojsisovics' zonal scheme continued to appear in text and reference books for many years. It is even in the current Edition of the Encyclopedia Britannica. The data from North America would be required to prove that the zonal scheme was not merely questionable but positively wrong (Chapter XV).

Carl Diener remained active in Trias studies until his death in 1928. In 1912 he summarized the Trias of the Himalayas, bringing together the data from all the expeditions and providing information and sketches from von Krafft's diary. In 1915 he published "Die Marinen Reiche der Triasperiode", a worldwide review. Although the emphasis in this paper is more on paleogeography than biochronology, it includes zonal schemes for both the Himalayas and the Alps. Mojsisovics' Juvavian zones are there, but they are now Norian. In his last years Diener wrote some important papers on the stratigraphy and ammonoid faunas of the Hallstatt Limestone. Collecting had not stopped with the completion of Mojsisovics monograph in 1902. Over the years new ammonoids were found by Kittl, Dr. August Heinrich and Diener himself. Heinrich, I believe, was a doctor of medicine, not geology. Most were from the famous Feuerkogel locality. Two publications from this period are particularly important regarding the Triassic time scale. One is "Leitfossilien", published in 1925. The other is Diener's very last paper, published in 1926, describing, as best he could, the exact situation of all the ammonoid beds and lenses in the Hallstatt Limestone.[199] The punch line of this paper, or, one might say, Diener's last words, was "I know of only two localities in Salzkammergut, Millibrunnkogel and Feuerkogel, where ammonoid beds can be observed in sequence. Mojsisovics only knew about the first". Millibrunnkogel was probably also the locality known to Suess and discussed by Dittmar (Chapter VII). Mojsisovics knew of several Carnian and Norian faunas from Millibrunnkogel. It is hard to know what Mojsisovics really did think about the succession. As mentioned in Chapter VII, he never realised his ambition to publish a detailed geological account of the Hallstatt area. Diener wrote that Mojsisovics, wounded by Bittner's hounding (Chapter VIII), eventually lost interest in the area that he had studied for so many years. If Mojsisovics did write an account, its whereabouts is unknown. This is because his widow sold his diaries and papers after he died. According to Diener they were sold abroad but I have not discovered where. But in any event the Millibrunnkogel locality can only be described as disastrously misleading. Carnian ammonoids appeared to be higher than Norian. Diener thought that this had greatly influenced Mojsisovics, many years before, when he concluded that Norian was older than Carnian. Diener thought that the beds at Millibrunnkogel had been turned upside down tectonically and that the failure to recognize this had caused Mojsisovics to make his mistake. In the 1970s it would be shown by Leopold Krystyn and his colleagues that Diener's interpretation was also wrong. The Millibrunnkogel "beds" do not form a sequence but instead owe their relative position to a variety of sedimentological complexities (Chapter XVI).

Diener's last words are a revelation: the final admission that Mojsisovics had no correct stratigraphic evidence for nearly the whole of his Upper Triassic zonation. Perhaps Diener withheld this until the end, motivated by a sense of loyalty to his old friend. The zonal scheme in "Leitfossilien" gives Diener's interpretation. Comparison with Mojsisovics' scheme shows one zone (*Heinrichites paulckei*) added; three (*Discophyllites patens, Cladiscites ruber, Sirenites argonautae*) subtracted. The Paulckei Zone was to accomodate the so-called "Mischfauna", one of the new discoveries at the Feuerkogel. This was the only place in all Salzkammergut where three zones could be seen in sequence (Aonoides, Subbullatus, Paulckei). This sequence was right, although today it is known that additional zones are intercalated. Diener's scheme still retained one of Mojsisovics' mistakes. The *Sagenites giebeli* Zone, for which Diener admitted he had no stratigraphy, is wrongly placed in the sequence.

XIII

1930s to 1950s: Spath and his influence

Leonard Frank Spath is generally thought of as a Briton, and so he was eventually. When Spath died in 1957, his biographer knew when he was born (1888), but not where.[200] Even his own family did not know. Only in the last few years has his son discovered that Spath was born in Germany and later moved to England, where he became a naturalized British subject.[201] He was educated at the University of London, mostly by attending evening courses. He gained a B.Sc. in 1912, and a D.Sc. in 1921. Eventually he became a Fellow of the Royal Society.

Spath's knowledge of cephalopod literature was encyclopedic. Nautiloids, Ammonoids, Coleoids; he seemed to have read everything and remembered it all. From 1912, except for a break during the 1914-1918 War, when he was a private in the British Army, he spent all his working life at the British Museum of Natural History in London. This museum houses the best collection of fossil cephalopods in the world. Spath seemed to know about every specimen. Early on he established a reputation for being so knowledgeable that collections were sent to him from all over the world, for identification, interpretation and description. Handling these collections added to his knowledge. Spath was never a permanent member of the Museum staff. He was offered a job in 1919 but on terms he considered unfair, in view of his age and experience. Spath's decision to avoid full-time employment may have been influenced by the difficulties of his family life. His wife was epileptic, a son was paralyzed. He cared for them both. He lived near the Museum and would generally spend the morning there, returning to his home in the afternoon to look after his family and write his papers. He owned a very comprehensive private library. At times he did some teaching but most of his income came as payment for reports he had written. He published extensively in the Danish publication, "Meddelelser om Grønland" and also in "Palaeontologica Indica". Payment was by the page. Unkind critics have suggested that the proliferation of genera and species that characterizes much of Spath's work was to produce more pages and thus more money. Spath wrote much on ammonoids and also a lot on biochronology but virtually nothing on the actual rock sequences in which the fossils were found. It has been suggested that Spath worked only in the Museum and never studied the fossils in their rock habitat. As far as the Trias is concerned this was apparently true. Living in England and seldom going abroad, he had no opportunity, there being no Trias with ammonoids in Britain. But according to Albert Reeley, a Museum employee who was his devoted assistant for many years, Spath was a thorough, careful collector in the Jurassic and Cretaceous rocks that were closer to home.[202]

Spath had a very keen eye for detecting subtle differences in ammonoid morphology. He was also convinced that ammonoid faunas were exactly similar only if they were exactly the same age. If they were different, they were of a different age. He was usually right, but not always right when he made tables showing faunas in sequence when he lacked geological evidence to show the relative age. He, like Buckman, believed that an exceedingly refined system of biochronology could be derived from elucidating the sequence of ammonoid faunas. Buckman and Spath sometimes had good ideas, but both lost sight of the paramount necessity for establishing the sequence in stratified rock.

Spath's major works that concern us are on the Lower Trias of East Greenland and Spitsbergen. Also, a complete review of all the Triassic genera entitled "The Ammonoidea of the Trias", published by the British Museum in two volumes (1934, 1951). These are known as Part IV and Part V of the Museum Catalogue Series. In Part IV Spath proposed a biochronology for the whole Trias that would prove very influential for about 35 years. Spath had very few new stratigraphic data. Some were provided by the geologists who had collected the specimens he described from East Greenland and Spitsbergen. Mostly he relied on the literature. His chronology was expressed in Buckman nomenclature – Otoceratan started the Trias; it ended with the Rhaetitan, and so forth. Some of his ideas were good. He was probably the first to appreciate the length of time represented by Lower Trias. But some of his ages were figments of his imagination; Carnitian, as the middle age of the Carnian, for example. Arthaber took him to task on that one.[203] Spath also believed that the Rhaetian represented a long span of time. I met Spath only once. It was about 1954. I was a beginner and much in awe. He was gracious and friendly, much more so than I expected, because his pen was often caustic. His sight was failing then. We chatted awhile and he gave me autographed copies of his East Greenland papers. On his desk was one of the old monographs with pictures of the Rhaetian ammonoid, *Choristoceras*. Rhaetian, he felt sure, was a long and important interval of geological time about which little was known. At the time I bowed to his judgement but I now think he was wrong (Chapter III). Spath's knowledge, judgement, and general astuteness deserve respect. In the 30 odd years since we met I have seen a lot more of the Trias. Too bad I cannot test myself against him now. When the time comes I would enjoy a debate with Mojsisovics, Diener and Spath in the Triassic heaven, if there is such a place that would admit us all.

On the whole I judge that Spath's influence in matters of Triassic biochronology was bad. This influence was exaggerated and prolonged from the effect that it had on Bernhard (Bernie) Kummel (1919-1980).[204] This has to do with the "Treatise on Invertebrate Paleontology". The Treatise project was conceived by Raymond C. Moore (1892-1974)[205] as a multivolume successor to the works published under the impetus of Professor Zittel, the famous paleontologist of Munich. Zittel textbooks on paleontology went through several editions, in both German and English, after their start in the 1890s. Moore was determined to enlist collaboration by experts from all over the world. By about 1950 he had experts for the Jurassic and Cretaceous ammonoids (W.J.Arkell and C.W.Wright). For the Trias he approached Spath. Spath's response: he would do a job; but it would have to be all the ammonoids and he would have to be paid. Trias alone, no; nothing, for no money. I do not judge him harshly for this. His ammonoid expertise had always been his livelihood. He was already past 60 and without any prospect of a decent pension. Moore had no funds to pay his authors. Arkell and Wright were already lined up. Clearly Spath's terms could not be met. So it came about that Bernie Kummel took on the job.

Bernie knew of Spath and of the British Museum collections. He had started Triassic work in about 1940 as a graduate student at the University of Wisconsin under the guidance of Norman Newell. The work they did together, and later efforts by Bernie on his own, led to a great improvement in the Lower Triassic faunal record in and around Idaho. Together they found *Claraia* near the base of the sequence, proving older Trias than known before. Even more important was Bernie's discovery of a very late Lower Triassic *Prohungarites* fauna above the *Columbites* beds. In making this discovery Bernie was the first to find a Lower Triassic fauna actually above the *Columbites* beds, the beds that J.P.Smith had thought were youngest in the division. This must have pleased Spath because Bernie had discovered stratigraphic evidence which confirmed a relationship that he had guessed. In 1951, just before taking up his job as a Harvard professor, Bernie spent nine months at the British Museum, talking to Spath and studying the collections. Spath's Part V was about to be published, and Bernie had access to the proofs. Bernie purchased Spath's library on Triassic ammonoids and brought it back to the United States. In some ways it is a unique collection. When Spath prepared "Part IV", he included, as text figures, illustrations from the classical monographs. To provide printer's copy, Spath had carefully cut the illustrations from the monographs. Later, with equal care, he restored them to their original position in the books. On some he had inked or whitened suture lines and chambers, for emphasis. Spath may have been a commercial artist at one time. It is on record that his drawings of suture lines were done by eye, without the control of any optical device.[202]

The classification of the Trias ammonoids that appeared in the Treatise (1957) is almost word-for-word Spath. This was acknowledged by Bernie, and it reveals the extent of Spath's influence. The Triassic chronology given in the treatise by Kummel regrettably incorporates the worst features of both Mojsisovics' and Spath's schemes. This is the one now in the Encyclopedia Britannica ![206] The combined efforts of Spath and Kummel towards improving Triassic biochronology were not a success.

AMERICANS,
A CANADIAN AND A BRITON

S.W. Muller
(1900 – 1970)

F.H. McLearn
(1885 – 1964)

L.F. Spath
(1888 – 1957)

B. Kummel
(1919 – 1980)

XIV

News from British Columbia and Nevada: Frank McLearn and Si Muller

During the years that Spath was meditating on Triassic biochronology in the British Museum, sections that would eventually provide many of the real answers were being found in North America, by F.H. McLearn in British Columbia, and S.W. Muller in Nevada.

First, McLearn. Frank Harris McLearn was born in Halifax, Nova Scotia, in 1885.[207] He was educated at Dalhousie University, later at Yale. There, as already mentioned, he was a student under Charles Schuchert. In 1917 he gained his Ph.D. with a dissertation on the Silurian stratigraphy and faunas of Arisaig. From 1913 until he died in 1964 McLearn was associated with the Geological Survey of Canada. Official retirement came in 1952 but he continued to work until the end. When he joined, the Survey needed a worker on Mesozoic paleontology and stratigraphy, so although his first research was on the Silurian, in the future it would be mainly on the Mesozoic. It was a gigantic job for one man. He not only did Triassic, but Jurassic and Cretaceous as well. The Triassic, he told me, was his favourite. He did field work throughout much of Western Canada. He put in about 25 field seasons; generally of three or four months. Only six brought him contact with the Trias. Nearly all the others involved younger rocks.

Trias had been discovered on Peace River, in the Foothills of the Rocky Mountains, by Selwyn in 1875 (Chapter VI). McLearn first studied Trias on the Peace in 1917, again in 1920 and 1922. On these occasions he had no opportunity to do detailed work, but found enough to realise that there was a section covering much of the Upper Triassic, with ammonoids at several levels. Fifteen years would pass before he would be allowed to return to these, his favourite rocks. Many of the intervening years were spent working on the Cretaceous and Tertiary of southern Saskatchewan. In 1935 and 1936 he had started what was planned as a fairly long job, on the Cretaceous of Manitoba. This was about 1500 km from the Triassic outcrops holding the ammonoids he had been yearning to collect for such a long time. McLearn was a quiet, somewhat reticent man, and his predilection for the Trias was not known to his official superiors, who probably did not care, anyway. He was also far from aggressive, and when it came to field work he did what he was told, and did not press his case for resuming Trias work. McLearn had started the Manitoba work with a younger colleague. The Chief Geologist of the time, G.A.Young, decided that he could make best use of his staff by having some of the younger workers take over the Manitoba operation. Young, it seems, did not quite know how to handle the transfer. He had the reputation for being a bit of a tartar, but he respected McLearn, and did not want to offend him. What happened next was told to me by Loris Russell, a friend and associate of McLearn's. Young came to McLearn's office and asked if he would agree to give up direction of the Manitoba operation, with the younger people taking over. Young clearly felt he was asking a favour, because with the transfer McLearn would have nothing to show, in the form of publications and so forth, for the couple of years work he had already done. Would McLearn, Young

NORTHEASTERN BRITISH COLUMBIA

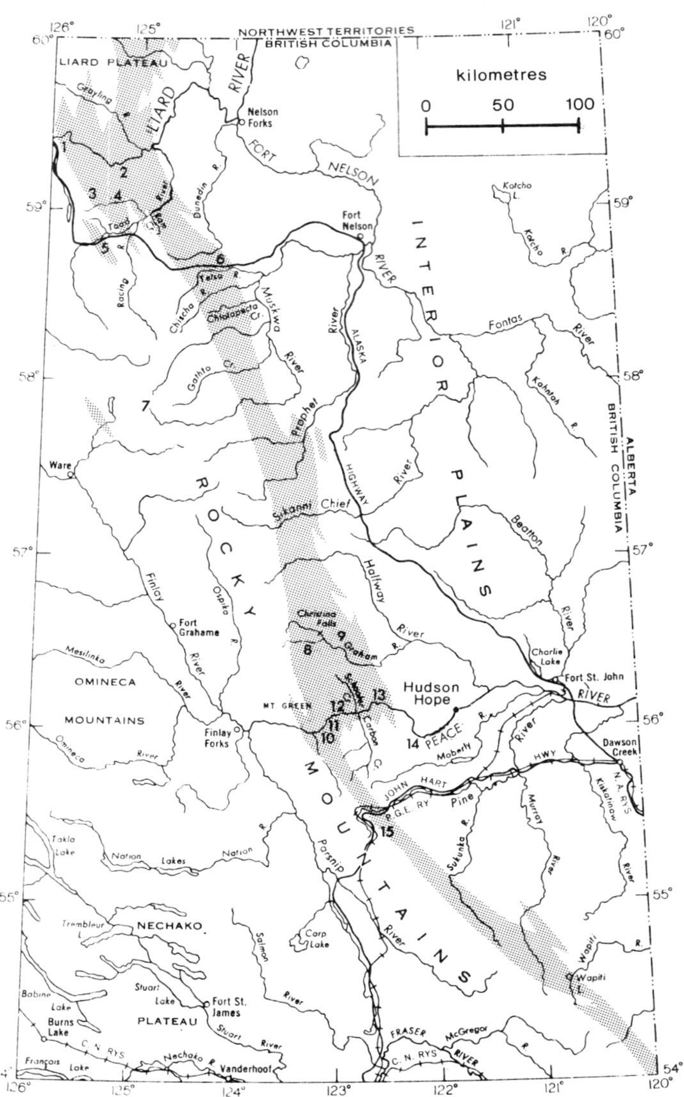

This shows Peace River before 1967. Today there is a large lake — Williston Lake — extending from the W.A.C. Bennett Dam (14), through the Rocky Mountains and up the valleys of the Finlay and Parsnip Rivers. Patterned area is mostly Triassic rock with some strips of Paleozoic, Jurassic and Cretaceous. Localities mentioned in the text: 1. Devil's Portage; 2. Hell Gate; 3. Mount McLearn; 4. Ewe Mountain; 5. Mile Post 428; 6. Mile Post 375; 7. Bedaux Pass; 8. Mount Ludington; 9. Crying Girl Prairie; 10. Ne-Parle-Pas Rapids; 11. Pardonet Hill; 12. Brown Hill; 13. Beattie Ranch; 14. W.A.C. Bennett Dam; 15. Pine Pass

went on, consider going back to study the Trias on Peace River instead ? Consider ! McLearn could hardly contain himself. Needless to say, he agreed to the proposition. No sooner had Young left, McLearn was in Russell's office, next door, to tell him that his dearest hope, one that he had entertained for 15 years, had suddenly and quite unexpectedly been realized. So it came about that McLearn was able to spend two long field seasons, in 1937 and 1938, working solely on the Trias of Peace River. This relieved Frank of a fear that had been nagging him all through 1936. That year Frank had been exchanging letters with Si Muller, at Stanford. Si wrote that he might make a trip to Peace River in the summer of 1936.[208] Frank was already booked to do Cretaceous in Manitoba in 1936. He had no immediate expectation of going to the Trias on the Peace. The thought of Si going to see the Peace River Trias, when he could not go himself, was very worrying. Early in 1937, this worry was dispelled.

McLearn's early work on the Peace, of which he gave an account in 1930, had shown that there was a section covering much of the Upper Triassic, with ammonoids from several levels. In 1937 and 1938 he made careful, bed-by-bed collections, particularly from three localities: Brown Hill, Pardonet Hill and Black Bear Ridge. He found at least a dozen ammonoid genera, previously known only from the Alps, the Himalayas or Timor. Up to then the only well known Upper Triassic ammonoid fauna in North America was the one with *Tropites*, described by J.P.Smith in California. Nothing to suggest this fauna was found, except for a few specimens at Pardonet Hill. That seemed strange. Brown Hill produced an interesting sequence: going up, *Stikinoceras, Malayites, Juvavites, Cyrtopleurites, Himavatites, Monotis*. Pardonet Hill had some that were the same; some were missing; some different, thus: *Stikinoceras, Tropites, Styrites, Himavatites, Monotis*.[209] Black Bear Ridge was mainly a mine for the *Himavatites* fauna. McLearn now had two problems. First, how did the Brown Hill sequence relate to that of Pardonet Hill. He concluded that they could be amalgamated as follows: *Stikinoceras, Tropites, Styrites, Juvavites, Cyrtopleurites, Himavatites, Monotis*. *Malayites* seemed to have a long range. Sad to say, his interpretation was wrong. It is now known that a major fault zone runs through Pardonet Hill. The *Styrites* and *Stikinoceras* faunas are about the same age (Kerri Zone, Table II). The *Tropites* fauna is the oldest of the lot. The fault is fairly obvious high on the hill, much less so low down. McLearn did all his work on the lower slopes. He told me that this kept him busy for about six weeks. The higher parts of the hill are steep and wooded, no pleasure to climb. His collections at the Geological Survey are testimony to his industry. When I went there in 1964 I found very little in the belt that he had traversed. It seems that he collected everything. He had been at every outcrop and demolished all the talus blocks. His notebooks give great detail; the position of every outrcop was carefully and accurately located; fossils from separate blocks were kept apart. So although he made a tragic error in failing to recognize the fault, and this led him to reconstruct a false sequence, his work is an absolute model, because all the data he acquired were available for later interpretation. McLearn's second problem was how to correlate with the Alpine sequence. This problem was a bit like the one that had faced Hyatt in California in the 1890s (Chapter X). It now hardly bears thinking about ! McLearn was trying to correlate his own sequence (which was wrong), with the Alpine sequence; this was also wrong. But he did not know this. *Tropites* indicated Carnian, *Cyrtopleurites* and *Monotis*, Norian. This he recognized. But he did not realise that the *Stikinoceras* and *Styrites* faunas were nearly the same age. The true position of *Malayites* had been established at Brown Hill but the significance of this discovery was obscured by the ambiguities at Pardonet Hill. McLearn's careful collecting at Brown Hill had established an incomparably better record for the Lower and Middle Norian than any other in the world. The discovery of *Tropites* on Pardonet Hill, recorded in his 1938 note book with delight and satisfaction, led him to conclude that the *Stikinoceras* level was

PEACE RIVER TRIAS – I

Above: Beattie Ranch on Peace River, about 1937.
F.H. McLearn photo.

Below: The same place in 1980, flooded by the waters of Williston Lake. Rocks are Ladinian.

PEACE RIVER TRIAS – II

F.H. McLearn's photo of his Brown Hill Section (Ladinian to Norian) 1938.
Below, new exposures formed on Williston Lake shoreline, 1980.

PEACE RIVER TRIAS – III

Above: Ne-Parle-Pas Rapids, 1964; Norian and Jurassic rocks, now submerged.

Below: Pardonet Hill, about 1937. Carnian and Norian, repeated by faulting. F.H. McLearn photo.

Carnian. But this *Tropites* discovery only confused the issue. If McLearn had gone only to Brown Hill and Black Bear Ridge, and had never found the Pardonet Hill *Tropites*, he would probably have got the right answer. Furthermore, if, instead of going to Pardonet Hill, he had joined Si Muller in Nevada, where Si had *Tropites* below *Stikinoceras*, and ammonoids that showed the affinity of the *Stikinoceras* and *Styrites* faunas, I believe that between the two of them, they would have straightened out the Upper Triassic zonation nearly 50 years ago.

McLearn also worked on Lower and Middle Triassic faunas. Much more is known about them now than in his day. One must remember that in the 1920s and 1930s nearly all of northeastern British Columbia except the Peace River valley was very inaccessible. There were no roads. The sternwheeler "D.A.Thomas" brought travellers up to Hudson Hope, below the Peace River Canyon.[210] This was quite unnavigable and the biggest obstacle for west bound travellers. It was the most formidable portage encountered by Alexander Mackenzie on his exploratory journey from eastern Canada to the Pacific Coast in 1793. In McLearn's day there was a wagon road that went from Hudson Hope, around the Canyon, to the Beattie Ranch, some 30 km to the west. Above the ranch the river was navigable for small boats for the whole of the Triassic belt, providing easy access. The Ne-Parle-Pas Rapids, above Pardonet Hill, could be exciting, but were easily managed on the north side of the river.[211] Ledges with *Monotis* projected on the south side. McLearn stayed at the ranch with the Beattie family, laying his sleeping bag on the sun porch, with a view of the Triassic hills. From the hill behind the ranch the Rockies formed the horizon. James Walker Beattie and his wife Elizabeth, who were pioneer settlers in the valley, became his good friends. James died in 1949 but his widow is alive today. Within a kilometre of the ranch house, at the river, was the Beattie Ledge, from which McLearn collected hundreds of exquisite ammonoids of the *Nathorstites* fauna. Originally thought to be Carnian, McLearn provided the evidence that most of the occurrences are Ladinian. The rocks at Beattie Ledge are the oldest Triassic exposed in the Peace River Valley. Further upstream, accessible by boats navigated by Beattie and his sons, were Brown and Pardonet Hills, and Black Bear Ridge, also the ledges at Ne-Parle-Pas rapids. These localities provided the Upper Triassic sections. In the Peace River Valley nothing older than Ladinian was exposed for McLearn's study.

Between McLearn's earlier and later visits to Peace River not much had been done on the Triassic there. In 1929 and 1930 M.Y.Williams and J.B.Bocock did geological work in the Peace River valley and made some Triassic collections in connection with surveys for the proposed Pacific Great Eastern Railway (now the British Columbia Railway).[212] Merton Yarwood Williams (1883-1974)[213] worked for the Geological Survey on and off for many years between 1912 and 1943, at first as a full-time employee. In 1921 he became a professor at the University of British Columbia but continued to work for the Survey in the summers. In 1930, Charles Mortram Sternberg (1885-1981),[214] the dinosaur collector and vertebrate paleontologist, was in the Peace River valley to study the Cretaceous dinosaur tracks in the Peace River Canyon and to collect Triassic ichthyosaurs. He made some good ammonoid collections at Beattie Ledge.

J.B.Bocock, Williams' younger colleague, presumably traversed the whole Triassic belt in 1934, in his capacity as second-in-command of the "Bedaux Sub-Arctic Citroen Expedition".[215] If Bocock made any geological observations or collections I am not aware that they have been recorded. The expedition, however, was quite an event. It was financed, to the tune of more than $250,000, by Charles Eugene Bedaux (1887-1943), with a contribution of $600 from the British Columbia government; they also contributed F.C.Swannell, one of their topographers. Bedaux was born in France but later became an American citizen. He had made a fortune by devising "the most completely exhausting,

PEACE RIVER TRIAS – IV

Franz Tatzreiter and Leo Krystyn photograph a Norian ammonoid. Near Childerhose Cove, Williston Lake.

The ammonoid. Lines are glacial striae.

inhuman efficiency systems ever invented";[216] his principle, evidently, was piece-work payment in industry, carried to an extreme. His expedition planned to cross the Canadian Cordillera with five Citroen half track vehicles. The route was explored and trails were cut by advance parties travelling with pack horses. Bob Beattie, the brother of James, was the head packer. Bob died in 1982. No less than 130 horses were used. The vehicles did not even reach the mountains but the pack horse parties crossed the Rockies between the headwaters of Muskwa and Kwadacha rivers, and a small party made it all the way to Telegraph Creek on Stikine River. The pass across the Rockies now bears the name of Bedaux; Fern Lake, in the pass, was named for his wife. She was on the expedition with two other women, one being Fern's Spanish maid. The main party travelled in style, with silverware, and the best of French food and drink. Bedaux himself crossed the Rockies but did not go much farther. They must have passed in sight of some outstanding Triassic sections in the Muskwa Valley, but as far as I can determine they did not avail themselves of the opportunity to collect the ammonoids that abound there.

Three years later Bedaux was again in the news, this time as the host, at Chateau de Candé, near Tours, France, for the wedding of H.R.H. the Duke of Windsor and Wallis Warfield, who was to become his Duchess but not a Royal Highness.[217] After abdicating as King Edward VIII in December 1936, the Duke, as he then became, had to choose a locale for the wedding ceremony. An opportunist, in the form of Bedaux, appeared on the scene. He offered his sumptuously appointed and renovated Chateau. The offer was accepted, surprisingly, because at the time he was not even acquainted with the Duke, let alone an old friend. After the wedding Bedaux cultivated the Duke and arranged for the Windsors to visit Nazi government officials in Germany. The Germans hoped to find an ally in the Duke. Bedaux, as go-between, ingratiated himself with the Nazis towards furthering his own industrial interests in Germany. His plans eventually backfired. When war came and France was defeated in 1940, Bedaux became associated with the Vichy government. In October 1942 he went to Algiers. His timing was bad. The Anglo-American-Free French invasion ("Operation Torch") took place on November 8. Bedaux was arrested. As a United States Citizen he was liable to be regarded as an American traitor. In December 1943 he was brought to the United States. There, about to be indicted for treason, he took an overdose of sleeping pills in Miami on February 18, 1944. To this day the nature of Bedaux's politics arouses controversy and his role as a Nazi sympathizer has recently been questioned. But he seems to have made friends on his British Columbia expedition. It was during the Great Depression and the men were more than pleased with their wages – four dollars a day. He was generous with his provisions. At one stage, far away in the British Columbia wilderness, everybody took part in a sumptuous feast of caviar, cheese and champagne.

Northeastern British Columbia has changed greatly since the 1930s. In 1942 the United States Army Engineers built the Alaska Highway, 2,400 km long, from Dawson Creek in British Columbia, to Fairbanks, Alaska.[218] The pioneer road was completed between March and November 1942. The route was almost complete wilderness, much of it mountainous. Access to country previously known only to a few trappers and Indians became relatively easy. The Geological Survey did not waste any time before taking advantage of the situation. In 1943 parties traversed the highway and much of the adjacent country. They brought back important collections of Triassic ammonoids. Particularly good collections were made by Edward Darwin (Ed) Kindle (1906-1982).[219] He was one of the sons of Edward Martin Kindle (1869-1940),[220] who worked for the United States Survey and later for the Canadian. Kindle Senior, in his U.S.G.S. days, had done work on the Triassic of Alaska. Ed made a boat journey on the Fort Nelson, Liard and Toad rivers, and found beautiful Lower and Middle Triassic ammonoids. Ed brought back a particularly spectacular specimen from the Anisian of Liard River. It was a *Sturia*, the first ever

STURIA

A MIDDLE TRIASSIC AMMONOID FIRST FOUND IN THE ALPS, NAMED FOR THE AUSTRIAN GEOLOGIST DIONYS STUR

This specimen is from Liard River, British Columbia

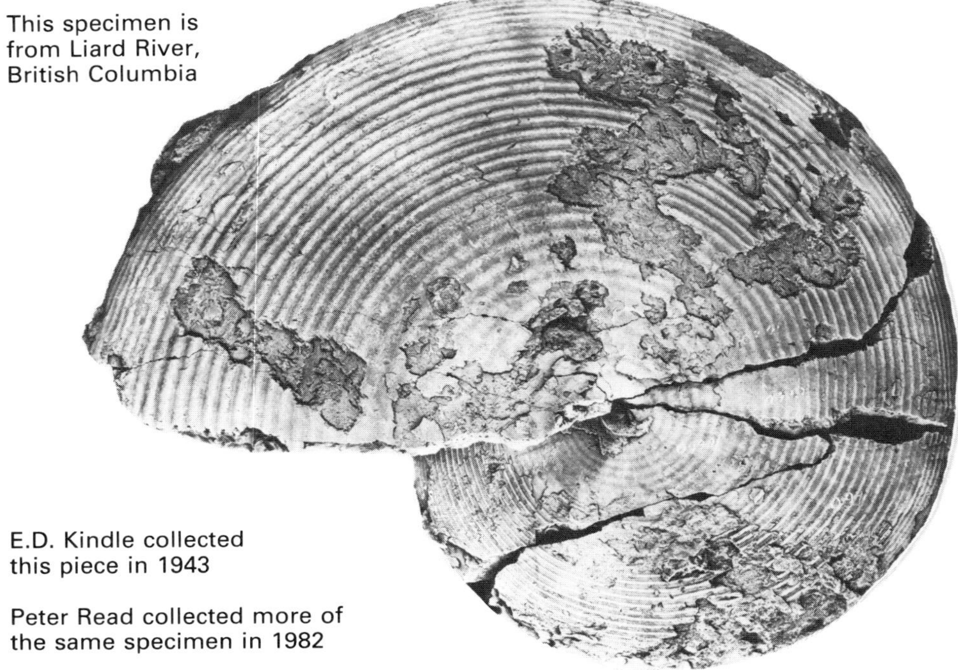

E.D. Kindle collected this piece in 1943

Peter Read collected more of the same specimen in 1982

M.K. Vincent shows them fitted together

found in North America. The specimen he brought back had a diameter of about 250mm. Although large, it was entirely chambered, and therefore not complete. In 1960 I was at Ed's locality and saw a large piece of a *Sturia*. I thought it might be a part of the same specimen but did not collect it because I was afraid it would break into small pieces. The locality was again visited in 1982, by Peter Read, who was doing geological work for the British Columbia Hydro and Power Authority. He successfully broke out the large piece, and sent it to Ottawa, where it was fitted to Ed's specimen. I was there again in 1983 and am happy to say that the exact stratigraphic position is still demonstrable, the impression of the ammonoid being clearly preserved on the outcrop.

Important discoveries around the Alaska Highway were also made in 1943 by Conrad O. (Con) Hage and M.Y.Williams. Besides working on the Highway Con made a trip with pack horses up the Sikanni Chief River which led to the discovery of important Middle and Upper Triassic localities. "M.Y." made the first reconnaissance of the Triassic rocks that intersect the Highway. The work done in 1943 led to the first discovery of ammonoids older than those at the Peace River localities. Anisian faunas were particularly abundant. *Nathorstites* also turned up, not surprisingly, because it had been found by R.G.McConnell on Liard River in 1887 (Chapter VI). When he saw all these specimens McLearn decided to go up there himself. This time he had no difficulty getting permission. The importance of studying sedimentary rocks and fossils was now appreciated by the Survey authorities. It had not been so in the 1930s. In 1944 Frank spent the whole summer in British Columbia. He travelled up the Halfway and Sikanni Chief River valleys with a pack horse outfit and worked on the Alaska Highway with a truck driven by his assistants. Here there are numerous fossiliferous Triassic outcrops between what for many years were Mile Posts 375 and 428 and as such are entrenched in the geological literature. They have now been metrified to 595 and 680, respectively. The important locality at Mile Post 428 was first spotted by Lowell R. Laudon, who travelled up the Alaska Highway with a party of students from the University of Wisconsin in 1946. On Sikanni Chief River McLearn was particularly pleased to find a section with the *Nathorstites* beds above the Anisian. Now he could relate the northern sequence to his Peace River sections. In 1966 I visited the Sikanni Chief River locality and looked at his section. I did not find many fossils. I found one big excavation, with a few scraps. It must have been Frank's quarry. Characteristically, his collecting had been exhaustive. He nearly always made very large collections. It was an embarassement de richesse, particularly for the Anisian. In places the Alaska Highway was literally littered with well preserved Anisian ammonoids. He did not have time to work out the detailed biostratigraphy. In his reports he thought that he was dealing with a single fauna. Later about six levels were found to be represented. Although McLearn did not know this, his notes and maps were carefully prepared, in such a way that all his data could be interpreted in the light of later discoveries.

Oil company geologists also sent in collections for identification, starting in 1944. Patrick K. (Pat) Sutherland, now a professor at the University of Oklahoma, was one of the first. In later years the companies were the first to use helicopters for geological work in British Columbia, leading to the discovery of the section at Ewe Mountain and many others. By 1960 the enormous extent of the northeastern British Columbia Trias really became apparent.

Great changes have also affected the Peace River valley. The rocky Peace River Canyon was obviously an ideal dam site. In the early 1960s construction was started. In 1967 it was complete. A gigantic reservoir, Williston Lake, was formed. The Beattie Ranch and ledge, and the *Monotis* beds at Ne-Parle-Pas Rapids, were submerged to a depth of about 200 m. James Beattie had died in 1949 and so did not live to see this happen. In 1981 his 87-year old widow Elizabeth was cheerfully living in Hudson Hope, having philosophically accepted the change. For the first few years the shoreline was a mess of tangled

TRIAS AND TRAVEL – NORTHEASTERN BRITISH COLUMBIA

Sternwheeler "D.A. Thomas" on Peace River below Hudson Hope, about 1920.

Jetboat at the lower end of the Toad River canyons, 1983. Rocks are Lower and Middle Triassic.

River boat "Seagull" with barge at Fort Nelson, 1960. This boat, without the barge, ascended Liard River to Hell Gate.

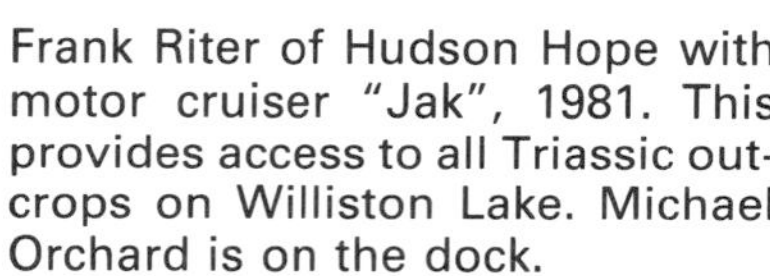

Frank Riter of Hudson Hope with motor cruiser "Jak", 1981. This provides access to all Triassic outcrops on Williston Lake. Michael Orchard is on the dock.

Geological Survey party travelling with pack horses in the Foothills north of Halfway River, 1966. The ridges expose Triassic rocks.

driftwood but it is gradually clearing. Now there are superb new shoreline exposures at all of McLearn's old localities and at the site of the Ne-Parle-Pas Rapids. I have visited them over the past few years and found new exposures of all McLearn's fossil beds with two exceptions: the *Nathorstites* beds of Beattie Ledge, and his *Tropites* bed at Pardonet Hill. I could not find the *Tropites* bed in 1964, when Pardonet Hill was much the same as when he was there. Nor could we find it in 1981, after the water level had risen.

McLearn was a bachelor. He lived with his sister Katherine (Kay) for most of his life. She was his chauffeur; Frank himself never drove a car. He travelled very little, except to the field. Buckman and Spath he knew only through correspondence because he never went to Europe. His U.S. counterpart was John B. Reeside Jr. (1889-1958). For many years they were acquainted only through the mails. Eventually, in 1947, they met at a Geological Society of America Meeting in Ottawa. Happily McLearn's achievements were recognized by the award of gold medals from both the Royal Society of Canada (Miller Medal, 1947) and the U.S. National Academy of Sciences (Thompson Medal, 1948).

McLearn died the year I went to Pardonet Hill and found the fault in his section. When I returned from the field he was in hospital for a prostate operation. There was every expectation that things would go well. I visited him there. He was rather depressed because Kay, his life-long companion, had died a few weeks before. I did not know what to say to him. He knew I had been to Pardonet Hill since we had last met. What should I say ? Should I tell him about the fault ? I had sought advice from a mutual friend and Geological Survey colleague, Robert John Wilson (Bob) Douglas, whose untimely death in 1979, was a great loss to geological science. Bob said that Frank would want to know. I told him. He never left the hospital alive. I have always wondered if I did the right thing.

Siemon W. Muller (1900-1970),[221] the other principal for this chapter, was born in Blagoveshchensk, a town on the Amur River, some 800 km northwest of Vladivostock. His father had come from Denmark to work on the trans-Siberian telegraph line; later he settled in the country as a teacher. Muller, later known to his wide circle of friends as "Si", was a student at the Russian Naval Academy in Vladivostock at the time of the Bolshevik Revolution in 1917. He managed to escape to Shanghai. Here he worked for an American company and began to learn English. A brother also escaped, but I believe that his father was murdered early in the revolution. In 1921 Si came to the United States where he took up geology at the University of Oregon. Having gained a degree there, in 1927 he went to Stanford to study under James Perrin Smith – "JP". This introduced him to the Trias. After doing his graduate work he stayed on as a teacher, eventually becoming a professor. Si had a very busy life. Besides teaching at Stanford he also worked for the U.S. Geological Survey and did a great deal of mapping in the structurally complicated Mesozoic rocks of Nevada. This was done with a team of collaborators. They all became good friends. Si's most notable association was with Henry Gardiner Ferguson – known to his friends as Ferg (or Fergie). Ferg worked for the U.S. Survey for many years. He died, aged 84, in 1966. Norm Silberling knew Ferg well. He tells me that there was an occasion when a dedicated bureaucrat, presumably looking for square pegs in round holes, circulated a request that members of the Survey describe how, in their own estimation, they were suited for the job they were doing. Ferg's reply was short and to the point: "by virtue of an indolent disposition and an independent income".

Around 1943 the U.S.forces needed to know more about permanently frozen ground, in connection with their Alaskan activities. Most of the literature on the subject was in Russian. Si, being a geologist with fluent Russian, was taken on to advise. He prepared a manual in which he coined the term "permafrost". So besides his notable achievements in stratigraphic and structural geology, Si left his mark in a totally unrelated field. Permafrost is certainly a word in our language today, although shortly after it was introduced a Harvard

Professor, Kirk Bryan, condemned it as an "etymological monstrosity".[222] He also complained that it too closely resembled names that had been given to brassieres and refrigerators. Bryan proposed "pergelisol" as an alternative. Si's reaction to this proposal was recounted to me by Norm Silberling. Norm was Si's field assistant at about that time. One day they were in town, and Si suggested that Norm go to the drug store complaining of an intestinal ailment, and ask for a bottle of pergelisol ! But Bryan did leave his mark because nowadays pergélisol has been taken into the French language as the word for permafrost.

Si became a thoroughly competent structural geologist, stratigrapher and paleontologist. He did so much field work and made such enormous collections that he never had time to describe more than a small fraction of what he had collected. He made notes and identified specimens. He knew the literature on Triassic and Jurassic ammonoids and bivalves intimately, and made excellent identifications. His expertise was recognized by his election, in 1965, to the Presidency of the Paleontological Society. In holding this office, in one respect he must be unique. In his whole career Si never published a formal, illustrated description of a single fossil ! I doubt if this could be said of any other President. There were descriptions and figures in his Ph. D. thesis, but it was never published. Triassic ammonoids of many different ages; Jurassic, too; *Monotis* and Jurassic Pectens; he collected and studied them all. But it was too much for the lifetime of one man and he seemed unable to focus long enough on any one of these interesting things. None of his paleontological projects was fully completed. However the fossil identifications he recorded in brief abstracts, marginal notes on maps and so forth are testimony to his industry and provide the measure of his achievements.

In 1936 and 1939, with Ferg as co-author, Si published two important papers on the Mesozoic stratigraphy of Nevada.[223] As a source of basic stratigraphic data for understanding the biochronology of the Norian, and the Triassic-Jurassic boundary, they are absolute classics. Si had started work in the Hawthorne and Tonopah quadrangles, southwestern Nevada, in 1927. He worked on through the 1930s. At meetings of the Geological Society of America he would announce his latest findings. These are recorded in the abstracts; very short, none as much as a page in length, but immensely significant nonetheless. In 1930 he reported finding the Norian *Pinacoceras metternichi* and basal Jurassic faunas in the Gabbs Valley Range.

In 1931 he recorded his discovery of an Upper Triassic ammonoid fauna, hitherto known only from Sicily. He had found this at West Union Canyon and it was of the greatest importance. The fauna in question had been described by Gaetano Giorgio Gemmellaro (1832-1904), of Palermo.[224] Gemmellaro's monograph gives descriptions of ammonoids from several different localities but there are no details or discussions concerning their age and sequence. Gemmellaro died before the report went to the press. Possibly more information would have been provided if he had lived a little longer. All the Sicilian ammonoids were Upper Triassic and some were similar to forms in the Hallstatt Limestone. Others, notably *Mojsisovicsites*, were new and different. Without stratigraphy the biochronological significance of *Mojsisovicsites* was completely unknown. Si found this ammonoid associated with *Stikinoceras* and *Guembelites*, at a level above *Tropites*. This paved the way for the recognition of the Kerri Zone (Table II).

By 1934 Si had found *Choristoceras* between the *Pinacoceras metternichi* fauna and the basal Jurassic. *Choristoceras* characterizes the Rhaetian, formerly regarded as a division of stage rank, terminating the Triassic. Si's section provided a sequence of ammonoid faunas covering the Triassic-Jurassic boundary far better than any known elsewhere in the world. This was true when Si found it in the 1930s; it is still true today. There may have been a short break in sedimentation between the Triassic and Jurassic but the section in the Gabbs Valley Range section comes closer to providing a continuous record of ammonoid faunas for this interval of time than any other. There is no foundation for the assertion that the topmost Triassic is missing in North America.[225] For assessing the significance of the Triassic-Jurassic boundary the Nevada section is of paramount importance. Better than any other, it places the Rhaetian in its true perspective.

Much of the work that Si did with Ferg was in the Hawthorne and Tonopah quadrangles, but he also did a lot in other areas. He worked all over Nevada, and also knew the Triassic geology of California, Oregon and Idaho. In 1937 he published a table recognizing a sequence of 18 faunas, ranging in age from earliest Trias (*Claraia* beds) to the very latest (with *Choristoceras*).

Muller's recognition of *Claraia* was important. It provided, for the first time, evidence that very early Triassic beds (now Griesbachian, Table II) were present in North America. Curiously *Claraia* had been collected in the Alberta Foothills, as early as 1906 by Donaldson Bogart Dowling (1858-1925), of the Geological Survey. The specimens had been sent to Professor Schuchert for identification, rather surprisingly because he made no claims as a Mesozoic expert. Schuchert misidentified them as *Monotis circularis*.[226] Nearly 40 years would elapse before *Claraia* would be recognized in Canada (Chapter XV).

In Si's 1937 table, with one exception, all the sequential relations were based on observations that he had made himself in the field. For the Carnian the malignant influence of Spath's meditations is apparent. A *Carnites* fauna is intercalated between the *Trachyceras* and *Tropites* faunas. From *Tropites* up Si had sections; he also had sections from *Trachyceras* down. But he did not have a section with *Tropites* above *Trachyceras*. His *Carnites* was misidentified and in any event had not been found between *Trachyceras* and *Tropites*. But virtually everything else was right. This table shows the astonishing extent of Si's achievements by 1937. There was more to come. JP knew the Middle Triassic mainly from the Humboldt Range where he coined the *Daonella dubia* Zone, which he thought to be mostly or wholly Anisian. Si found new Middle Triassic localities in the Tobin Range and at Augusta Mountain. Francis Newlands Johnston, another of JP's students at Stanford, made important discoveries in the New Pass Range. Between them they showed that the *Daonella dubia* Zone had at least two Anisian levels (*Acrochordiceras* followed by *Paraceratites*), with *Nevadites* and *Protrachyceras* beds above. This confirmed the Alpine Anisian zonation (Binodosus Zone, followed by Trinodosus Zone, Table II, p. 8). This example of Alpine zonation proved to be one of Mojsisovics' better guesses ! The relationship in Nevada is much clearer than in the Alps. The ammonoids are in dark-coloured bedded limestones in which the stratigraphy is clear. Many of the occurrences with comparable ammonoids in Europe are in the red Hallstatt limestone facies where the determination of exact sequence is plagued by the ever present problems of condensation. *Protrachyceras*[227] is now known to be above *Nevadites* and is the best marker for the base of the Ladinian. The important thing revealed by this work was that the Anisian-Ladinian ammonoid sequence in Nevada was incomparably better than any known in the Alpine-Mediterranean areas.

Clearly the work of McLearn and Muller had provided enough stratigraphic data to provide a Triassic biochronology vastly superior to that of the Alps. The magnitude of their contributions was not appreciated; neither by themselves nor by others. Both used the Alps as a standard. Mojsisovics' Norian scheme was accepted lock, stock and barrel. The hints that it had grievous shortcomings, apparent from the writings of Kittl and Arthaber, and eventually also Diener, were not taken. They would have done better to have ignored it; also better to have dismissed the scheme generated by Spath's meditations. They had the data to construct their own chronology. Why didn't they ? As far as McLearn was concerned it was perhaps irreconcilable with his temperament. He was modest to a fault and bowed to the authority of Mojsisovics and Spath. Si, as I knew him, was not of this temperament. He was

a better linguist than McLearn and could easily read all the European literature. Also he travelled more. In 1937 and 1938 he was in Austria looking at Museum specimens and collecting fossils in Salzkammergut and elsewhere. Being such an experienced field geologist I am surprised that he did not appreciate that the stratigraphic foundation for nearly all of Mojsisovics' zonal scheme was not merely tenuous, but nonexistent. Mclearn and Muller never met. They exchanged letters from 1933 on, swapping specimens of *Stikinoceras* and *Mojsisovicsites* and discussing their significance. In January 1936 Si suggested visiting Frank in the field so that they could go over the Peace River outcrops together. On April 22 Si wrote "My plans for this summer are not yet definitely worked out, but should the time permit I will make an effort to visit some of the Triassic localities in Canada". We have already seen that this worried Frank because it was before he had been given the green light to resume his own work on Peace River. He was not keen on the idea that Si might go there first, by himself. To Frank's relief, Si did not make the trip in 1936. In March 1937 Si writes that he will be going to Europe in the summer and will call at Ottawa on the way.[228] By June 5, 1937 Frank had the green light and wrote to Si: "This summer I am returning to the Peace River field to extend my stratigraphic studies and to make more collections. I am hoping, but not guaranteed that this is part of a two year program. It would be very desirable to me, if after your summer in European museums you could go over the sections with me". This never came about. Si never made it to Peace River. He stopped in Ottawa on his way from California to Europe. Frank had left for the field but had set out his material for Si to see. But they never acutally met one another. Had they done so perhaps they would have anticipated many of the results that accrued from the collaboration of their successors, 20 years later. Si's successor was to be Norman Silberling; I was to be Frank's. It's worth remembering that Si had found *Stikinoceras* (with *Mojsisovicsites* and *Guembelites*) above *Tropites* and that Frank seemed to have them the other way round. It was a problem, because at that time Nevada and Pardonet Hill were the only places in the world where any relationship was known. If they could have met and talked it over perhaps things would have been reconciled, Pardonet Hill would have been rejected, the full significance of *Stikinoceras* would have been appreciated and the whole concept of the Norian would have been clarified. *Stikinoceras*, it would turn out, characterizes the Kerri Zone (Table II), a Norian interval that was completely unknown to Mojsisovics, Gemmellaro and Diener. That is why Frank and Si, using the old chronology, could not express its age. Emancipation from the Alpine standard was clearly necessary.

XV

Emancipation: An Independent American Time Scale

This chapter is mostly about work done in the United States and Canada over the past 30 years by Norman J. Silberling and the writer. By 1967 this led to the introduction of a biochronological standard based entirely on North American data; a standard completely emancipated from that of the Alps (Table II, page 8).

I joined the Geological Survey of Canada in 1952, the year Frank McLearn retired. He was still coming to the office every day, driven in the car by his sister, and actively revising and amplifying, for final publication, the preliminary accounts that he had published over the preceding twenty years or so. The early reports had been published in the Transactions of the Royal Society of Canada and the Canadian Field-Naturalist. Later they were issued by the Geological Survey in the Paper Series, which at that time were mimeographed texts, with printed plates of fossils, all enclosed in a manilla envelope. Now that Frank had retired it was his intention to prepare final, properly printed publications. By 1960 his final report on the Pardonet faunas was published. When he died he left an almost complete manuscript on the Anisian. I prepared this for the printer and it appeared in 1969. Nearly all the specimens and data described in these reports were acquired by 1944, the last year that he was in the field.

When I arrived my main job was to work on clams and snails; things that lived in freshwater and on land in Cretacous and Paleocene time. Not very exciting fossils, compared with ammonoids. There was not a lot to do. The Survey had not replaced Frank with another Trias worker. It was decided that I would gradually take over, under his guidance. He would continue to work on the collections from northeastern British Columbia; I was expected to work elsewhere and to identify collections submitted by other members of the Geological Survey. I was certainly fortunate to have Frank as a teacher and mentor.

I started Trias field work in 1953, in the southern Yukon, around Lake Laberge, where, in the words of the poet Robert Service, they cremated Sam McGee. We travelled on the lake and Yukon River in canoes with outboard motors. The sternwheel steamer "Whitehorse" was still in service between Whitehorse and Dawson. At intervals the steamer would drop off supplies for our field party. This is the area where G.M.Dawson had collected corals and other fossils in 1887, but which were not recognized to be Triassic until the work of E.J.Lees (Chapter VI). Before I went there interest in these Triassic rocks had been aroused by the work of John G.Fyles and J.O.Wheeler. They had worked in the Whitehorse area, just south of Lake Laberge. John Fyles had started the work in 1946, nominally under the direction of W.E.Cockfield. He had the misfortune to catch tuberculosis from the camp cook, and was out of geological action for about three years in

YUKON TRIAS AND TRANSPORTATION

The sternwheeler "Whitehorse" coming up the Yukon River on the way to Lake Laberge (below). Distant hills on the river and on the east (left) side of the lake are Upper Triassic limestone. 1953.

consequence. Happily he did not share the fate of Gabb (Chapter VI). John's brother, Jim (J.T.) Fyles has also done extensive geological work in the Cordillera, in various capacities, eventually as a Deputy Minister, for the British Columbia government. Although most of his work has been on rocks other than Trias, in the 1950s he did detailed mapping around Cowichan Lake, a very important Triassic locality soon to be mentioned. In 1982, when with Peter Read, he participated in the collection of the second piece of the only good *Sturia* known from North America (Chapter XIV).

John Wheeler took over from John Fyles in the Whitehorse area in 1948. He is the third generation member of the well-known family of Canadian Alpinists, all sharing the middle name Oliver. The first two generations were distinguished topographers. John became a geologist. John's father, Edward O. (eventually Sir Oliver), worked for the Indian government for many years; in 1921 he was a member of the First British Expedition to Mount Everest. His grandfather, Arthur O., did extensive Survey work in the Rockies and Selkirks. Arthur married Clara, the daughter of John Macoun, who had been with Selwyn on Peace River in 1875 (Chapter VI). So John Macoun, a former Assistant Director of the Geological Survey, was John's great grandfather.

Both Johns (Fyles and Wheeler) made extensive Triassic collections. It eventually turned out that the Triassic of this part of the Yukon was of considerable significance, both for biochronology and tectonics. For biochronology it confirmed placing the Amoenum Zone above the Cordilleranus Zone. Tectonically the rocks belong to what is now known as a "Suspect terrane". The rocks contain what may be described as a warm-water fauna, despite their present high latitude. All the evidence suggests that they have been displaced northwards by 1000 km or so, in the time that has elapsed since their deposition.

At the end of the summer I went south via the trail of '98, over the White Pass to Skagway on the narrow gauge railway; thence to Vancouver by ship, calling at Alaska ports on the way. Steam locomotives were still in use on the railway. It was a romantic, leisurely, way to travel.

In Vancouver I met George (Jurij Alexandrovich) Jeletzky, a Geological Survey colleague. Since 1949 George had been doing very detailed stratigraphic work on the west coast of Vancouver Island. His main job was Cretaceous, but he was investigating coast lines that had never been mapped in detail, and while pursuing Cretaceous, also encountered very important localities for the Triassic, Jurassic and Tertiary. It was now my job to identify the Triassic fossils that George had collected. He had found all sorts of interesting things, in particular beds with a big variety of fossils, including heteromorph ammonoids, in beds above those with *Monotis subcircularis*. In terms of the present chronology, George was the first to demonstrate Amoenum Zone above Cordilleranus Zone. My Yukon results seemed to agree with George's. Some of his fossils suggested a correlation with the Sutton Limestone, which had been discovered at Cowichan Lake, on Vancouver Island, by C.H. Clapp, of the Survey, back in about 1910. Clapp's fossils had been described in a joint paper with Hervey Woodburn Shimer (1872-1965)[229] with the surprising title "The Sutton Jurassic .. of Vancouver Island". Shimer was a professor at the Massachusetts Institute of Technology who occasionally worked on material from Canada. There is a lesson to be learned from the title of this paper ! Shimer identified one ammonoid in the fauna – *Choristoceras*. He knew that this was Triassic in the Alps. But he also identified some corals and clams which he compared with Jurassic forms. He had one Triassic ammonoid; several others kinds of fossils that he thought to be Jurassic. He counted heads and took the vote to be for the Jurassic! Not the way to do biochronology. For biochronology ammonoids are in a class by themselves. When present they cast the deciding vote. All others are disenfranchised. Dating by counting is sometimes done today, glossed with an veneer of sophistication by using statistics. Statistics, albeit crude, misled Shimer. They may be misleading well intentioned biochronologists today. My field companion, would, I think,

A TRIASSIC FOSSIL NAMED FOR A SHIP . . .

Daonella frami
found at Blaa Mountain . . .

by Per Schei (1875 – 1902) . . .

. . . geologist of the Second Norwegian Polar Expedition in the "Fram" (1898 – 1902) led by Captain Otto Sverdrup.

. . . AND A SHIP NAMED FOR A FOSSIL

A species of *Halobia* from Griesbach Creek, Axel Heiberg Island

The Canadian ship "Halobia II"

under power . . .

at harbour . . .

and under sail.

agree.[230] Together George and I went to Cowichan Lake, and there, thanks in large measure to his industry, we obtained a good collection of ammonoids – *Choristoceras, Rhabdoceras* etc. This fauna I now think is youngest Triassic – Crickmayi Zone. The only unfortunate thing is that there are no accurately dated beds above or below the Sutton. Thus the position of this important fauna in the sequence is not demonstrable.

It was originally planned that I would continue to work in the Yukon. But in the winter of 1953-1954 I came into close contact with another Geological Survey colleague, Raymond (Hromundur) Thorsteinsson. Ray and I had been graduate students together at Toronto in 1949. He left Toronto that year having decided to continue at the University of Kansas, where Raymond C. Moore held the fort. More significant towards eventual progress in Trias studies, was Ray's (Thorsteinsson, not Moore) own introduction, in 1950, to work in the Canadian Arctic Islands. By 1953 he had already spent four seasons up there and had discovered immensely important Paleozoic sections. He also knew well the literature on the early explorations. We would talk and reflect on the possible extent of Trias in the islands. Occurrences on Ellesmere Island were known from the work of Per Schei (1875-1905),[231] the energetic geologist of the "Fram Expedition" (1898-1902) and Johannes Troelsen, who went there in 1940 with a Danish Expedition. Schei had found Triassic fossils at several places in Eureka Sound, between Axel Heiberg and Ellesmere Islands, namely at Ammonite Mountain, Hat Island and Blaa Mountain. These fossils were described by Ernst Kittl, of Vienna. In August 1953 Alf Erling Porsild (1901-1977) had collected specimens of *Halobia* from eastern Axel Heiberg Island, adding evidence that Triassic rocks might be extensively exposed in the Eureka Sound area. Porsild was the distinguished botanist and Arctic expert. He was Danish by birth, had been brought up in Greenland, and eventually became a Canadian. On this occasion he was a member of a private expedition sponsored by George Jacobsen of Montreal. The party landed on a lake, now known as Buchanan Lake, in a Canso flying boat. The lake was chosen as a possible landing site because it had been spotted to be free of ice in late summer by the Arctic geographer, Diana Rowley, from her inspection of the trimetrogon air photographs. The pilot was Welland W. ("Weldy") Phipps, the navigator Jock Buchanan, and the party, besides Jacobsen and Porsild, included Diana's husband Graham and Dalton Muir. They landed on the lake, where Porsild collected the fossils. They also walked to the west and were the first to reach the ice cap that covers the centre of Axel Heiberg Island. It was the first time that an aircraft had landed on the interior of Axel Heiberg Island. From seeing the high Arctic tundra on this occasion Weldy Phipps first conceived the idea of landing light aircraft on this kind of terrain, without the necessity of a prepared airstrip. As we will see, he was later to develop this technique and it was put to extensive use by parties of the Geological Survey. Jock Buchanan was lost shortly afterwards, apparently by drowning when trying to save the same Canso when it was dragging its anchor in a Scottish loch. His body was never found. He is commemorated in the name of the lake. The lake had been seen before, ice-covered, by the Danish Expedition in 1940. They had named it "Maersk Lake" for the shipping company, one of their sponsors. The Canadian government authorities decided on Buchanan Lake in preference.

By 1954 the Arctic Islands, although still pretty remote, had become a little more accessible through the establishment of weather stations. The Geological Survey of Canada, ever ready to work on the frontiers of the nation, took advantage of the situation by using the weather stations as take-off points for field work. Ray Thorsteinsson had worked from Resolute, the station on Cornwallis Island. He had travelled by canoe and on foot, on the first occasion with Yves Fortier and Trevor Harwood of the Defence Research Board. Bill Heywood had been to Isachsen; Bob Blackadar to Alert. Bob Christie was about to go to northern Ellesmere Island with Geoffrey Hattersley-Smith, also of the Defence Research

TRIASSIC OF THE CANADIAN ARCTIC ISLANDS

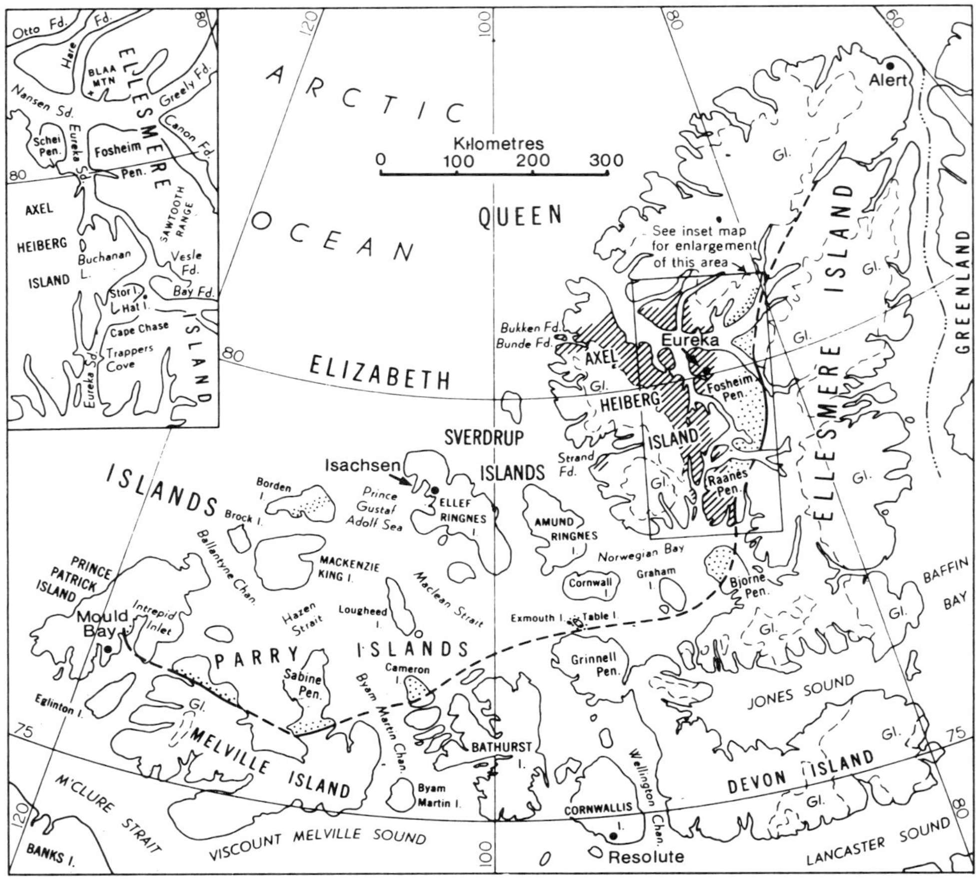

Pattern of dots shows where the Triassic rocks include much sandstone (Bjorne and Schei Point Formations). These rocks were deposited close to the shore line. Oblique lines show where the sediments were mainly mud and silt (Blind Fiord and Blaa Mountain Formations). They were deposited farther from the shore. The youngest Triassic Formation (Heiberg) is mostly sand and includes some deltaic deposits with coal. This was more uniformly developed throughout the area, compared with the older formations. The Heiberg is exposed within the patterned areas and also on Brock and Cornwall Islands. Drill holes have proved that fossiliferous Triassic rocks occur below the surface of Mackenzie King, Lougheed and Amund Ringnes Islands. There are no Triassic rocks south and east of the heavy line, which was probably close to the shore line.

Ammonite Mountain, where Per Schei collected Triassic fossils in 1901, is on Bjorne Peninsula. The creeks for which the Lower Triassic stages are named are in northern Axel Heiberg Island and northern Ellesmere Island. Griesbach Creek flows into Bunde Fiord; Diener, Smith and Spath Creeks are on the peninsula between Hare and Otto Fiords.

"Gl." shows the extent of glaciers.

Board. Bill's only means of transportation was his feet; the two Bobs had dog sledges. Nobody had been to Mould Bay, on Prince Patrick Island, or to Eureka, on Ellesmere. Ray suggested that I should go to Mould Bay, taking in by air, an Inuk from Resolute with his dog team to provide transportation.

Since prehistoric times there had been no permanent Inuit settlements in the high Arctic Islands. In the 1920s and 1930s a few Inuit from Greenland had wintered there as travelling companions of the Royal Canadian Mounted Police. In 1935 the Canadian Government moved a group from Hudson Strait to Dundas Harbour, Devon Island. After one winter there, they moved across Lancaster Sound to Arctic Bay, Baffin Island.[232] In 1953 the Canadian Government started another resettlement plan by bringing about a dozen Inuit families to Resolute from more southerly villages. One family, that of Amagualik, came from Pond Inlet, on northern Baffin Island, the others from Port Harrison (Inoucdjouac) in Hudson Bay. Amagualik and his family came with me to Mould Bay. We were later joined by Andrew Macpherson, a biologist, the brother of the poet, Jay. Andrew was young, but already had acquired several years of Arctic experience with the Arctic scientist and traveller Thomas Henry (Tom) Manning.

The expedition was successful for the Geological Survey. Sadly, for Amagualik, it was marred by the death of his five-year old daughter Zipporah. When we left Resolute on April 17 Zipporah was clearly not well but did not seem seriously ill. At Mould Bay her condition worsened. Medical advice was provided on the radio, drugs were given, but there was no improvement. It was decided to send an aircraft to take her to hospital. On the evening of April 30 a Lancaster of the Royal Canadian Air Force reached Mould Bay but owing to low cloud was unable to land. Zipporah died a few hours later. Despite his grief, Amagualik continued to work for me. Understandably at times his morale was low. He had lost his daughter and, when travelling, was, for weeks on end, without any Inuit for company and conversation. In future years we found it made for happier long journeys either to take two Inuit, or to have one Kabloona and one Inuk. Two garrulous Kabloonas tended to make a solitary Inuk feel particularly lonely and homesick.

Andrew brought a recorder for his and our entertainment. He was moderately but not perfectly proficient. I think it could be said that he was not possessed of perfect pitch. One of his pieces was the well known largo of Handel. This he would play, perfectly, except, to my ear, for one note. It sounded wrong to me, but I could not read the music sufficiently to suggest alternative fingering. It eventually turned out that my ear was better than his fingering !

From the Mould Bay weather station we travelled by sledge over the sea ice to various parts of Prince Patrick Island, and also to Eglinton and Melville islands. Summer comes late up there and we were still travelling on the ice in July. I had no certain expectation of finding Trias but was hoping for it. Permian and Jurassic I could expect, from the discoveries made by one of the officers of Sir Edward Belcher's expedition, referred to in Chapter VI.

The officer in question was Commander M'Clintock, Captain of the steam tender "Intrepid", one of the ships under Sir Edward's overall command. Eventually he became Admiral Sir Leopold M'Clintock, K.C.B., D.C.L., Ll. D., F.R.S. and Vice-President of the Royal Geographical Society. On three occasions he travelled in the Arctic Islands seeking traces of the Franklin Expedition. His Expedition of 1859, in the screw yacht "Fox", was successful. The earlier expeditions had been of the British Admiralty. The "Fox" Expedition was privately sponsored and organized. The ship was entitled to fly the White Ensign, M'Clintock being a member of the Royal Yacht Squadron, but he chose instead to fly the burgee of the Harwich Club, of which he was also a member.[233] A record recounting the death of Franklin was found in a cairn on King William Island. The record was written by officers who survived longer than Franklin, but everybody had perished by the time the record was found. M'Clintock was a very capable geological observer and fossil

ARCTIC TRIAS I – WINTER

Blaa Mountain, Ellesmere Island (1956).

Raanes Peninsula, Ellesmere Island, 1957. Steep cliffs are Cretaceous gabbro intruding Triassic shales.

Oodlatetuk on Exmouth Island (1957).

ARCTIC TRIAS II – SUMMER

Above: Hat Island in Eureka Sound, Upper Trias, R. Thorsteinsson and freighter canoe, 1956.

Below: Carboniferous to Lower Trias, Raanes Peninsula, Ellesmere Island. Soft basal Trias overlies hard dark Permian.

ARCTIC TRIAS III – SUMMER

Above: Folded Upper Trias, Stor Island in Eureka Sound, 1955.
Below: Horizontal Middle and Upper Trias, Hare Fiord, Ellesmere Island, P. Harker photo, 1963.

collector. He made a good collection in 1853 but had to leave most of it in the middle of Melville Island. He was returning from the exceptionally long sledge journey, more than 2200 km, on which he had discovered Prince Patrick Island. It was July. There was not enough snow for the sledge. The party then tried using a cart, but the ground was too soft. So they left almost everything with the cart and walked back to their ships which were frozen in the ice on the south coast of Melville Island. I say almost everything advisedly; M'Clintock brought back one Jurassic ammonoid, later named *Ammonites m'clintockei*, in his pocket. He left a bigger collection, and all his equipment, with the cart. He intended to go back, but was frustrated by circumstances. First, the ships were driven off the coast by a northerly gale, on August 16. It was therefore impossible to make a trip in the autumn. In the following spring Belcher gave orders for everybody to abandon their ships and return to England. So M'Clintock was never able to go back for his things. In 1958 Ray Thorsteinsson and I went to Melville Island and he found the cart, equipment and specimens. The ammonoids were carefully wrapped in paper: pages of the 1853 Nautical Almanack.

After 6 months of field work in 1954 I came away having found lots of interesting Paleozoic, Jurassic and Cretaceous, but, I thought, no Trias. I certainly had no Triassic ammonoids. It later turned out that I had in fact collected Triassic fossils: clams and brachiopods. These I had mistakenly thought were Jurassic. To have been sure of good Trias I should have gone to Eureka. I later asked Ray why he suggested I should go to Mould Bay and not Eureka. He admitted that he was being sly; he was saving Eureka for himself, because he knew that it was in an Arctic paradise, both climatically and geologically. I do not complain; on the contrary, when the time came for him to go to Eureka, in 1956, he invited me to go with him. Before that, in 1955, we were both in the Islands, on a big party with helicopter support,[234] led by Yves Fortier who was soon to become the Director of the Geological Survey. This led to acquisition of important Triassic collections by Brian Glenister, Digby McLaren, Jack McMillan, Jack Souther and myself. But the real bonanzas were revealed later, leading to the discovery of particularly good sequences of Lower Triassic faunas on Raanes and Svartfjeld peninsulas, Ellesmere Island and at Griesbach Creek, Axel Heiberg Island. These would eventually provide the grounds for establishing the Griesbachian, Dienerian, Smithian and Spathian stages. The stages were named for geographic features – creeks – according the the rules. The creeks had previously been named to commemorate workers, all of whom figure in earlier chapters, who had studied the Lower Triassic. These stages have now been adopted in many parts of the world. They have been sanctified to the extent of being included in the American Geological Institute glossary. There they are mistakenly said to be of European origin ![235] Many critical localities, I freely and gratefully acknowledge, were first discovered by my friend Ray Thorsteinsson. He was the first to find the localities at Smith and Spath creeks; at Griesbach Creek we were together; Diener Creek localities I found by myself. It was my pleasure, also, to visit Exmouth Island, by dog sledge in 1957 with the Inuk Oodlatetuk. Here I found Middle Triassic ammonoids associated with ichthyosaur bones like the ones that Sir Edward Belcher had brought back more than a century before. In all I spent 7 seasons on the Trias of the Arctic Islands between 1955 and 1964. Dogs and sledges were used until 1957. Since then both fixed and rotary winged aircraft have been used increasingly. In 1958 the Geological Survey started using Piper Super Cub aircraft fitted with oversize tires at low pressure to land on unprepared terrain. By then this form of transportation had been invented by Weldy Phipps.[236] We have seen that Phipps first conceived the idea when he piloted the Canso to Buchanan Lake in 1953. Aircraft equipped in this way could land within walking distance of nearly every outcrop in the Arctic Islands. The Arctic tundra became a delightful place to study the Trias.

To find such interesting Lower Trias in the Arctic was very satisfying, because it was a time interval for which British Columbia had contributed very little, so far. But there were

MOUNTAINS OF TRIAS – BRITISH COLUMBIA

Complete section of Carnian, Mount McLearn (above).
Anisian to Lower Carnian section on Ewe Mountain (below).
N.J. Silberling writes notes. 1965

some things common to the Arctic and British Columbia (e.g. *Claraia, Wasatchites.*) Obviously it would be worth trying to tie the sequences together. Towards this end I went to northeastern British Columbia in 1960, and since then have made about 10 further trips. In addition to my own collections, and those of Frank McLearn, I have been fortunate in having available those of Geological Survey colleagues, notably D.W. (Dave) Gibson, E.J.W. (Win) Irish, Jan E. Muller, Walter W. Nassichuk, B.R. (Bern) Pelletier, D.F. (Don) Stott, and Gordon C. Taylor; also collections made by geologists from many oil companies. Nearly all of McLearn's old localities have been reexamined, and many new ones found.

The Peace River and Halfway River localities are within an extensive tract mapped by Win Irish between 1959 and 1962. Win had previously spent about 15 years mapping in the Alberta Foothills, and this had led to important discoveries of Griesbachian, Smithian and Anisian. For the work north of Peace River Win's assistant, in 1960-61, was Walter Nassichuk. Walter was the first to find two important localities: Crying Girl Prairie and Mount Ludington, both of which contribute immensely important collections and data for interpreting the Norian. Walter was later to take up his own ammonoid studies, concentrating on the Carboniferous and Permian. Irish's work in Alberta and most of the work in British Columbia was done in the days when horses provided the principal means of transportation. There were few roads. Even now there are not many. There were some trails cut by the local hunters and trappers. For the work north of the Peace a second generation of the Beattie family played an important role. They were James' (Chapter XIV) three sons, Robert (Bob) and his younger twin brothers Jim (James Hudson) and Bill (William Hope). This Bob should not be confused with his uncle, who had been the packer for the Bedaux Expedition in 1934. The ranch on Peace River, at the end of the road from Hudson Hope, was often the base for the operations. From there pack horses would carry the camping gear, equipment and provisions. The geologists would ride when possible, but most of the work was done on foot. The Beattie brothers, as packers, would load and control the horses, which not infrequently were perverse creatures, clearly with minds and opinions of their own !

In 1958, Bern Pelletier and W. Blake Brady, also using horses, made a detailed geological map of the area that included McLearn's Alaska Highway localities, leading to the discovery of many new ones for the Anisian and Ladinian. Bern went on to make a special study of the Trias rocks; even more localities were brought to light.

From 1960 helicopters became the principal means of transportation. Dave Gibson continued Bern's work on the Triassic rocks; Gordon Taylor did the mapping. Don Stott made important discoveries in the upper part of the Trias sequence. This was the basement as far as he was concerned, because he was working mainly on the younger Mesozoic rocks.

Oil company geologists also made important discoveries. They were the first to find Lower Carnian, both south of Peace River and at Ewe Mountain. An oil company geologist who subsequently joined the Geological Survey was E. Wayne Bamber; he made some important Ladinian collections near the Muskwa River, the country traversed by the Bedaux Expedition (Chapter XIV).

I have been able to visit nearly all these localities myself but without the work of these colleagues it would have taken much longer to piece the story together. All this made possible the establishment, in northeastern British Columbia, of a sequence of standard zones from the base of the Anisian to the base of the Upper Norian.

An important result from Triassic studies made in the foothills and eastern ranges of the Cordillera over the past 30 years has been the realization that northeastern British Columbia represents only one segment of a belt (albeit discontinuous) of wholly sedimentary Triassic rock that extends all the way from the Alaska border south to the 49th Parallel, a belt some

TRIAS IN FOOTHILLS AND CANYONS BRITISH COLUMBIA

Perfect exposures of Ladinian. Foothills south of Chischa River and in the Grand Canyon of Liard River. (below)

2500 km long. This is what is known as the plate-bound Triassic, in contrast to that of the suspect terranes to the west. The first clues to the existence of this belt were provided by Selwyn and McConnell (Chapter VI). The next important clue was provided by another of the Geological Survey explorers, Joseph Keele (1863-1923). Keele discovered *Monotis* on Rackla River, in east central Yukon, in 1905. In 1907-1908 Keele made a notable exploration for the Survey by crossing the Mackenzie Mountains. He ascended the Ross River and descended to the Mackenzie on what was then known as Gravel River, now as Keele River. Some idea of the conditions under which he worked is provided by his own words: "During the journey we built two boats and one cabin, and, until the Mackenzie was reached, saw no person except a small band of wandering Indians at the head of Gravel River". The cabin was necessary because the party spent the winter on the divide. They travelled by boat when possible, but for about 100 miles on the divide this was impossible; this part of the journey was accomplished during the late winter, "the necessary outfit being hauled on sleds with three dogs, in relays. . .".[237] Keele's early discovery has been followed up by the findings of L.H. (Lew) Green, J.A. (George) Jeletzky, D.K. (Don) Norris, J.A. (Jim) Roddick and Dirk Tempelman-Kluit. The southern part of this great Triassic belt is in the Foothills and eastern Ranges of the Alberta Rocky Mountains. This came to be appreciated in 1945 when Percival Sydney Warren (1890-1970), who at times worked for the Geological Survey, and became a Professor at the University of Alberta, identified *Claraia* and Anisian ammonoids from the Rockies. One of Warren's students was C.R. (Charlie) Stelck, who eventually also became a Professor at the same university. Stelck's paleontological interests are manifold. He has not done a great deal on the Triassic, but is nevertheless interested and knowledgeable, having been one of McLearn's assistants on Peace River in the 1930s. He made an important contribution of his own by being the first to recognize Triassic ammonoids in the wells drilled for oil and gas on the Plains, east of the Foothills. G.B. (Geoff) Leech made a significant contribution by finding determinable Lower Triassic bivalves in the Fernie area, near the southernmost part of the sedimentary belt of Trias.

Most of my own work in western Canada has been in this great belt, where the rocks are wholly sedimentary and there are magnificent sections in the foothills and eastern ranges of the Rocky Mountains. As already mentioned, the sequences here, like those of the Arctic Islands, were deposited on the American Plate. They are folded and faulted but are in general amenable to stratigraphic interpretation. They provide an outstanding record for all the Trias except the earliest and latest.

On a few occasions I have ventured into the Western Cordillera, the belt of suspect terranes where the presence of volcanic rocks, structural complexity and less perfect exposures commonly make things much more difficult and ambiguous. For the biochronological standard, the most important area is Tyaughton Creek, in southern British Columbia, where I visited the Geological Survey party of Howard Tipper in 1953. This is an area where Colin Crickmay had collected very well preserved fossils, mostly Jurassic ammonoids, in 1939. He also collected *Monotis, Rhabdoceras* and *Choristoceras* and a distinctive Triassic clam named *Cassianella*. The sections are broken by many small faults. Crickmay had plotted the position of his finds with great accuracy, but he evidently did not have time to piece the section together. Armed with his locality data, and with a helicopter for transportation it did not take long to sort it out. This now provides the standard biochronology for the latest Triassic (Cordilleranus, Amoenum, Crickmayi zones), the interval that is not completely represented in northeastern British Columbia or in the Arctic. In this area *Cassianella* is very abundant in the Amoenum Zone. I hope that Crickmayi Zone stays in use, as testimony to Colin's achievements in Cordilleran geology. Certainly I am in his debt. The Tyaughton Creek section is more informative than Si Muller's in the Gabbs

LATEST TRIAS – WESTERN BRITISH COLUMBIA

George Jeletzky collecting Upper Norian ammonoids from the Sutton Limestone, Cowichan Lake, Vancouver Island, 1953.

Upper Triassic limestone, forming summit, is overlain by *Monotis* beds and *Cassianella* beds (Upper Norian, Cordilleranus and Amoenum Zones). North side of Tyaughton Creek Valley.

Cassianella beds (Amoenum Zone) are overlain by *Choristoceras* beds (Crickmayi Zone). Tyaughton Creek.

Valley Range for the Trias, but for the boundary with the Jurassic the Nevada section tells more. The two are complementary.

Also now known to be on suspect terranes are the rocks in southern British Columbia – the Vancouver and Nicola groups – which were recognized as Triassic by G.M.Dawson in the late 1800's (Chapter VI). Parties of the Geological Survey have done extensive work since then. The work of Clapp, Shimer and Jeletzky has already been mentioned. H. C.Gunning and Colin Crickmay did important work on Vancouver and the adjacent islands in the 1920s and 1930s. More recently work was continued for the Survey by J.W. (Jack) Hoadley and Jan E. Muller and by Donald Carlisle and Takeo Susuki of the University of California. Atholl Sutherland Brown did extensive work in another of G.M. Dawson's old areas, the Queen Charlotte Islands, for the British Columbia government. The Nicola rocks of south central British Columbia form an extensive terrane and related formations extend up the central part of the Cordillera all the way to the Yukon border. Much of the Nicola is of volcanic origin and less amenable to detailed stratigraphic study compared with the Vancouver Group, in which most of the Carnian and Norian is wholly sedimentary. Important data and collections have nevertheless been obtained from the Nicola and related rocks by Survey officers, before and during the 1939-1945 war by J.E. (Jack) Armstrong, Hugh S. Bostock, Clive Elmore Cairnes (1893-1954), William Egbert Cockfield (1890-1956), Clifford Symington (Cliff) Lord (1908-1981), and Harington Molesworth Anthony (Tony) Rice (1900-1970); also by Colin Crickmay, who did field work in many parts of British Columbia at his own expense. Stanley (Stan) Duffel and Kenneth C. (Ken) McTaggart also worked on the Nicola. Copper deposits are often associated with these rocks and they remain an object of study, contemporary workers for the Survey being R.B. (Dick) Campbell, J.W.H.(Jim) Monger, Peter B.Read and Howard W. Tipper. British Columbia government geologists – W.J. (Bill) McMillan and V.(Vic) Preto have also been active; also Mikkel Schau and William B.Travers of Cornell University. Virtually all the collections made by these people were sent to Ottawa, providing pieces to fit the puzzle.

Much of the Triassic of western British Columbia is in very rugged country. I have never visited the most rugged – the Coast Range around Stikine River. The relief is great. Kates Needle (3000 m) stands above the river, which is nearly at sea level. Timber line is around 1000 m, with dense Pacific rain forest on the lower slopes. The mountains are draped with glaciers. Forrest Alexander Kerr (1896-1938) worked in this country between 1926 and 1932. Kerr mostly used his legs to do the geology. Sometimes pack dogs helped by carrying his gear. The dense forest made it difficult to use pack horses. Horses and river boats provided the standard means of transportation in much of British Columbia in those days. Starting around 1956 helicopters came to be widely used. Until then the field work of the British Columbia geologists was extremely arduous as the rule, not the exception. Kerr collected Triassic ammonoids including some identified by McLearn as *Stikinoceras kerri*. This was both curious and interesting. This species was first discovered by McLearn on Peace River in 1922. Kerr was McLearn's assistant, and made the topographic map on which the Peace River localities are so accurately plotted. The fossil name commemorates Kerr, and alludes to an occurrence in the Stikine River area, but the type specimen is from the Peace. Kerr's pioneer work was followed up, with important additions, by Hubert (Hu) Gabrielse, and J.G. (Jack) Souther and most recently by Jim Monger and Howard Tipper. Again, more pieces for the puzzle.

Now is the time to introduce Norman Silberling (b. 1928). He was a student under Si Muller at Stanford. His very first field work was as Si's field assistant in Nevada, in 1948. Norm started his own field work on Trias with the United States Geological Survey in the early 1950s, when I was starting in Canada. First he worked in Nevada; later in Alaska, both on the North Slope and the Panhandle. His Nevada work provided the first grounds for

COLLEAGUES AND COLLABORATORS

K. Seyed-Emami and M. Davoudzadeh in the Anarak area, Iran.
Nearly flat Cretaceous limestones unconformably overlie nearly vertical Lower and Middle Triassic beds. 1972

With Vanda Kollarova–Andrusovova and Riccardo Assereto in the Kendelbach valley, Austria, near the place where O.C. Marsh discovered *Choristoceras* in the 1860s. 1973.

With Norman Silberling in British Columbia, 1965. . .

. . . and Raymond Thorsteinsson at Eureka weather station, Ellesmere Island, 1956.

subdividing the Upper Carnian. Later work, in the Humboldt Range, led to the elucidation of the best sequence of Spathian to early Ladinian ammonoid faunas known anywhere in the world.[238] Particularly important was his discovery that the Lower Triassic *Subcolumbites* beds, were followed first by the Haugi Zone (previously known only from California), then by the Lower Anisian Caurus Zone (Table II). This provided an unrivalled record for the Lower-Middle Triassic boundary. For a while Norm was a professor at Stanford, the successor to JP and Si. One of his students, Kathryn Nichols, did important work on the Anisian. Now they are husband and wife as well as scientific collaborators. I first met Norm in 1954. We swapped information then and have ever since. In 1964 we arranged to meet in Nevada. That year I saw quite a bit of Trias ! I was on holiday in Salzkammergut in the spring. I didn't achieve anything to speak of but it was interesting to drive past the Feuerkogel, and to take the cable cars up to Steinbergkogel and Siriuskogel. At Siriuskogel I saw specimens of *Halobia* in the walls of the Inn, but nothing in situ ! Later I went, first to the Arctic, then to Pardonet Hill, and then drove south to meet Norm in Nevada, stopping at JP's Lower Triassic localities in Idaho on the way. Si Muller joined the party and we made a memorable excursion to the Humboldt, Tobin, and New Pass ranges, also Augusta Mountain and the Inyo Range locality in California. By then I had read Diener's last paper (Chapter XII) and had seen Salzkammergut. We talked a lot. Clearly it was time to demolish the so-called Norian standard. While we were at it, why not try to make our own standard for the whole Trias, basing it solely on North American stratigraphic data that we could test ourselves. Forget about the Alps ! It had caused nothing but trouble. It had certainly led Frank and Si astray. We embarked on the project and arranged a second joint field excursion. This was to northeastern British Columbia in 1965. Both these trips were immensely rewarding. We added much to the Anisian zonation in both British Columbia and the Humboldt Range. Between us we had data from the length and breadth of the Cordillera and also the Arctic Islands for establishing a Triassic ammonoid biochronology. We had data from several hundred localities; sequential data from dozens. The Cordilleran belt is some 5000 km long and up to 1000 km wide. In terms of global areal coverage for the marine Trias, we were personally acquainted with about a third of it all ! It seemed worthwhile to construct a biochronology based solely on this substantial chunk of the world's Trias and then see how the rest of the world fitted it. We did it and established a chronology of some 34 zones in sequence.[239] We had caught up with von Krafft at last and were emancipated from the tyranny of the Alpine chronology !

At many levels Triassic ammonoids are remarkably cosmopolitan; *Wasatchites*, for example, clearly permits correlation of late Smithian rocks all over the world. At other levels, notably from the Anisian on, the faunas from the Arctic are very different from those in the Tethys, and correlation may be difficult or impossible. The faunas of northeastern British Columbia perform an important role in resolving some of these problems. They have affinities with those of the Arctic, and also with those of Nevada. Those in Nevada, in turn, can be related to those in Tethys. For this reason the North American chronology, since 1968, has proved particularly useful for worldwide application.

CHINESE AND RUSSIAN WORKERS

Liang Yi-luo, Yuan Yi-ping, E.T. Tozer, Zheng Zhuo-guan, W.W. Nassichuk, He Guo-xiong, Wang Yi-gang. Nanjing, 1978.

M.N. Vavilov, M.V. Korchinskya, Yu.V. Archipov, unidentified, L.D. Kiparisova, T.M. Astachova, Yu.N. Popov, T.M. Okuneva, A.N. Oleynikov, A.I. Zhamoida, Leningrad, 1966.

XVI

Worldwide Reconciliation

Turning away from North America, much has been achieved in other parts of the world in the last thirty years. Considerable effort has been spent restudying the Permian-Triassic boundary beds. More significant have been the results obtained from the main part of the Triassic succession, particularly since 1968, which have shown that the sequences in the Alps and elsewhere in the Tethys can be readily reconciled with those of North America.

First we will consider the Permian-Triassic boundary beds. At Dzhulfa and Julfa (respectively the Russian and Iranian localities on opposite side of the Arax River) it was known, from the early work of Abich, Frech, and Arthaber, that there were Permian beds with otoceratids. These are peculiar and distinctive ammonoids, obviously related to *Otoceras woodwardi*, which characterizes the basal Trias. Mojsisovics judged them more primitive and consequently older. That time he was evidently right, despite the dangerous nature of his argument. Also known from Dzhulfa was *Paratirolites*, an ammonite vaguely like *Tirolites*, of undoubted Triassic age. The genus was discovered and named by Alexander Stoyanow (1879-1970)[240] who studied the Djulfa section in the early years of the century, and who provided, in 1909, its first accurate description. Stoyanow was born in Russia but most of his professional career was in the United States. *Paratirolites* was originally regarded as Triassic, but for no good reason. *Claraia*, undoubtedly Triassic, was known above *Paratirolites*, from the work done by Stoyanow. From all these old data, Dzhulfa seemed a possible place for finding an uninterrupted marine sequence covering the boundary. A team of Russian paleontologists studied the section in detail and came to the conclusion that such a sequence was, in fact, preserved. But soon after their report was published, in 1965, Zhao Jingko (=Chao King-koo), writing from Nanjing, pointed out that their interpretation was almost certainly wrong, and that most of the beds thought to be lowermost Triassic were in fact Permian. Zhao's interpretation is now accepted by nearly everybody, including most of the Russians. Something that closely resembles *Paratirolites* is now known in the Permian of China, which seems to clinch the case.

The sequences in southeast China are of considerable interest in connection with the boundary. Back in 1939 Hsu Te-you identified a squashed ammonoid, collected near Nanjing, as an *Otoceras*. This suggested basal Triassic. Later Zhao and his colleagues discovered a remarkable assortment of Upper Permian ammonoids, seemingly as young or younger than any known elsewhere. Southeast China thus seemed a possible place to tell what might have happened between Permian and Trias. In 1978 a party of Canadian geologists, including me, led by Walter Nassichuk, went to China to look over the situation. We were taken to Hsu's locality but found no *Otoceras*. At the Nanjing Institute the specimen could not be located. The person most likely to know its whereabouts, Zhao, was ill, and not there. We moved on to other places without seeing it. Zhao, in hospital, was told we wanted to see it. He knew exactly where it was. It was found, and our gracious Chinese hosts arranged for the curator of the Museum to bring it to Guangzhou, for Walter and me to look at, a couple of days before we left China. This was pretty good service ! The round trip was about 2000 km. I looked at the specimen and thought that it was truly an *Otoceras*, although being squashed, I was not wholly confident. If it is, it means that China preserves a geological record known nowhere else in the world: youngest Permian followed by earliest Trias. But even here there is a very distinct bedding plane separating the two, and continuous sedimentation seems unlikely. Something that is awaiting settlement is the choice of a type locality for the Trias base. China seemed a worthy candidate if it really had *Otoceras*. Since our visit the Chinese have been back to the locality again, this time with a party of Japanese scientists. They now think that Hsu's ammonoid is something else – a *Koninckites*.[241] This would indicate an age appreciably later than basal Trias. All the ammonoids collected near the base of the Trias in southeast China that I have seen are squashed. It's probably best to consider them unsuitable for precise identification. The basal Trias stratotype must have good ammonoids, so China would not be a good choice.

Hsu devoted 12 years to hunting for Triassic ammonoids. His career ended when he was killed by bandits in Guizhou, in 1944. Rui Lin, of the Nanjing Insitute of Palaeontology, who was one of our hosts in 1978, has provided this account of the tragic incident. Dr. Hsu, and two other geologists, were examining Lower and Middle Triassic rocks in western Guizhou province. It is mountainous terrain and was known to be a favourite haunt of bandits who pillaged the villages. Outsiders were recognized as likely prey. Dr. Hsu and his colleagues had been shopping and obviously had money. This was noticed by the bandits. The geologists were followed to their hotel. Here the bandits saw their boxes for specimens, and assumed that they were full of money, not fossils and mineral specimens. The time came for Dr. Hsu to travel to a different area. It was necessary to go by foot, and engage porters. Alert to the possibilities, the bandits managed to get the job. Shortly after starting, the men that the geologists believed to be porters suddenly rushed ahead, then turned and fired on the geological party. Dr. Hsu was killed instantly. The other two were captured, but eventually they were also killed, in order to eliminate witnesses.

Japanese scientists have travelled widely to look at Permo-Triassic boundary beds. In 1969 a group led by K. Nakazawa and H.M.Kapoor of the Geological Survey of India discovered Trias with *Otoceras* above Permian at Guryul Ravine, near Srinagar in Kashmir. This relationship had been known in the Himalayas since Griesbach's day, but the Guryul Ravine succession was more accessible, and ideal for detailed study. Curt Teichert and Bernie Kummel had been there the year before but had not found *Otoceras*. Bernie told me that he was suffering from dysentery at the time of his visit, which was very bad luck because it hampered his field work and may have deprived him of the chance of making the fossil find. Much has now been written about this section, and I vote that it become the stratotypic base for the Trias, being endowed with well preserved *Otoceras* in a well documented, relatively accessible situation.

Waagen's Salt Range sections have also received a lot of attention in the last 30 odd years. Otto Schindewolf (1896-1971)[251] and Adolf Seilacher went there in 1952. They found an *Ophiceras*, which occurs with and just above *Otoceras* in the Himalayas, but they did not find *Otoceras* itself. Nobody else has either, so the existense of earliest Trias in the Salt Range is unproven and unlikely. Kummel and Teichert were there in 1961-1962; Jean Guex, with Aymon Baud and Louisette Zaninetti in 1975. All added important details, but I think it is fair to say that Waagen's pioneer work has stood up remarkably well. Jean's work led to the proposal of the Nammalian Stage for the middle part of the Lower Trias.[252]

Back in 1926 James Mann Wordie (1889-1962),[242] the British polar explorer and geologist, who was eventually knighted and became Master of St. John's College, Cambridge, had discovered Lower Triassic ammonoids at Kap Stosch, in East Greenland.[243]

Soon afterwards extensive collections were made by Danish expeditions led by Lauge Koch.

Lauge Koch (1892-1964)[244] was an important, if somewhat controversial figure. He was involved in the study of Greenland geology for nearly half a century, starting in 1913. The work in East Greenland, where the Trias occurs, was done between 1926 and 1958. Koch was an enthusiastic geologist. He was also a great organiser, efficient in raising funds for his expeditions and for arranging publication of the results in the "Meddeleser om Grønland". His expeditions were truly international. He engaged scientists from Britain, Switzerland, Germany and Sweden, as well as Denmark to work in the field and on the collections. Many scientists benefited, and enhanced their own reputations by participating in Lauge Koch's expeditions. Hans Frebold and Curt Teichert went there in the summer of 1931. Frebold returned in the autumn; Teichert spent the winter near Kap Stosch and continued his work in 1932. Augusto Gansser was a geological assistant in 1934, shortly before he undertook the Himalayan expedition with Arnold Heim (Chapter IX).

Most of the Triassic collections were made between 1927 and 1934 by Alfred Rosenkrantz (1898-1974) and Eigil Nielsen (1910-1968). Nielsen became a specialist on Triassic fish, but he made large collections of ammonoids, which occur in the same formation as the fishes. The ammonoids were sent to Spath at the British Museum. Spath's first major report appeared in 1930. It announced the discovery of *Otoceras*. This was exciting news. Until then *Otoceras* had been known only from the Himalayas (Chapter IX). The presence of earliest Trias in the Arctic was firmly established. The ammonoids from East Greenland are beautifully preserved. Unfortunately it had proved difficult to determine the exact stratigraphic position of many fossils. This is because the beds in which they occur are soft and tend to form mud which slides downhill on top of the permafrost. In consequence, many of the specimens were not collected exactly in place. Circumstances were not ideal for working out the exact sequence.

Lauge Koch himself was not directly involved in the Trias work to a great extent. Being so busy with organising and directing these large expeditions he had little opportunity. But even if less knowledgeable on the details than the experts he had conscripted, he took an enthusiastic interest in all aspects of the work. He retained this interest all his life. This was very clear when I met him at the International Symposium on Arctic geology held in Calgary in 1960.

In the 1930s Koch's enthusiasm led to an unfortunate incident. Feelings were generated and thoughts expressed reminiscent of the 1898 affair in Vienna (Chapter VIII). In 1935 Koch published a book on the geology of Greenland.[245] Inevitably it was not perfect and some of his generalisations were not justified. No less than 11 geologists banded together and wrote an unfavourable, indeed savage, review.[246] Five were people who had gained their expertise on Greenland geology as members of Koch's expeditions. Attention was focussed on some inaccuracies and misinterpretations. He was also accused of plagiarism. Koch considered that he had been libelled and took the matter to court.[247] The courts found the review within the law. But in another related case Koch was judged justified in his complaint.[248] So he won one and lost one.

The outcome of the court cases did not discourage or deter Koch from continuing his Greenland work. The 1939-1945 War caused interruption but expeditions were resumed between 1947 and 1958. A new international group was enlisted. Not surprisingly his antagonists of the 1930s were not involved. Rudolph Trümpy joined the 1958 Expedition and made a detailed study of the Lower Triassic beds. He made a strenuous effort to determine the exact position and sequence of the ammonoid beds.

The Lower Triassic rocks of East Greenland include beds of conglomerate. Some of the conglomerates contain fossils – brachiopods and bryozoans – that look the same as those in the underlying Permian formation. At one time Professor Trümpy thought that these fossils had actually lived in Triassic time and that East Greenland might provide a record of continuous sedimentation between the Permian and Trias.[249]

In 1967 Trümpy went back. Koch by now was dead. This time it was a party organized by Professor Tove Birkelund of the University of Copenhagen. Bernie Kummel and Curt Teichert were invited; also present was Eigil Nielsen. Teichert and Nielsen were returning to the area they had first visited 35 years before, as members of Lauge Koch's expedition. The party reexamined the conglomeratic beds and concluded that the fossils that looked to be Permian, although in Triassic rock, were in fact Permian. In other words, they were derived.[250] They also agreed that the boundary between the Permian and Triassic beds marks an interruption in sedimentation, as seems to be true wherever marine Trias and Permian are in sequence.

Now to review some of the discoveries in the younger Triassic beds.

Around 1908 a distinctive Lower Triassic ammonoid fauna characterized by *Subcolumbites* and *Albanites* had been discovered in Albania. The position of the fauna in the sequence was not known, however. Next it was recognized in Timor, but this was not much help. Thanks to the work of L.D. Kiparisova (near Vladivostock), Zhao Jingko (at Lingyun, Guangxi, China), H. Bender (at Chios), Bernie Kummel (in Afghanistan), M. Davoudzadeh and K. Seyed-Emami (in Iran) considerable data were acquired towards positioning this fauna in relation to the Lower-Middle Trias boundary. The most precise positioning, however, is provided by the sections worked out in the Humboldt Range, Nevada, Nevada by Norm Silberling.

A great deal of work has been done on the Middle and Upper Trias of Tethys. Nearly all the classical localities in Salzkammergut, the Dolomites and elsewhere in the Mediterranean region have been restudied. The anomalies that appeared when the North American sequence was compared with the supposed sequence in Europe have been resolved. There is no longer any doubt about the defects in Mojsisovics' chronology. Most of this work has and is being done by Riccardo Assereto of Milan, Giulio Pisa of Bologna, Leo Krystyn and Franz Tatzreiter of Vienna, Max Urlichs of Ludwigsburg, Volker Jacobshagen of Berlin, and Hans Rieber of Zurich, to name the ammonoid workers. All except Assereto and Pisa are still with us. Tragically, on September 15, 1976, they were both killed by a rock slide triggered by an earthquake. This was while they were doing field work on the Triassic of Monte Bivera, in the Friuli region of northern Italy. Riccardo's young son was with them, and also killed.

Much of the Tethys work has been done under the stimulus provided by Professor Helmuth Zapfe, of the University of Vienna, who organized a project on the Tethys Trias under the International Geological Correlation Programme. The meetings he organized enabled the new generation of Tethys workers to become acquainted with one another and their North American colleagues. Fruitful dialogue and friendships were the result.

Assereto clarified the position and relationships of the classical Anisian localities in Italy and in Kokaeli Peninsula, east of Istanbul. Anisian fossils had been described from this part of Turkey many years ago by the Vienna paleontologists, Franz Toula (1845-1920)[253] and Arthaber (Chapter XII), but virtually nothing was known of the faunal sequence until Riccardo went there. This, and work on the Greek Island, Chios, led him to propose a new classification for the Anisian. The Aegean Substage was proposed for the earliest part, followed by Bithynian and Pelsonian. I think the distinction between Bithynian and Pelsonian is of a low order, best expressed at the zone level. Riccardo and I used to argue about this. Chios is important, being the only place in the Mediterranean area where Spathian and Lower Anisian beds are exposed in sequence. It has been proposed as a stratotype for the Lower-Middle Triassic boundary but is not suitable because the ammonoid sequence is incomplete, the interval of the Haugi Zone being unrepresented.

Good ammonoid sequences for the Upper Anisian and early Ladinian are very rare in Tethys.The classical locality of Epidauros, in Greece, is currently being studied, from collections obtained by extensive quarrying, by Leo Krystyn and Franz Tatzreiter. Hans Rieber's work on the ammonoid and *Daonella* sequence in the Swiss Alps is also important in this connection. Some European authors favour placing the base of the Ladinian at a lower level than the one chosen in North America. They would prefer the Occidentalis Zone in the Ladinian. It would be good to have agreement here, but decisions should await full documentation of the Epidauros sequence. For world-wide application, the boundary position chosen in North America has much to recommend it. At the lower level it would be unrecognizable in the Arctic.

Urlichs has done much with the ammonoid faunas of the famous St. Cassian locality we've already heard so much about. From this we can expect improved understanding of the Lower Carnian chronology and the Ladinian-Carnian boundary.

Krystyn has also done an immense amount of valuable work towards untangling the Hallstatt limestone sequences at the Sommeraukogel and Feuerkogel. An ammonoid sequence at Sommeraukogel has been provided for the first time; the Feuerkogel record greatly improved. Both provided data for the Carnian and Norian which were readily reconciled with the sequence that had been found in North America. From Krystyn's work it also turned out that many of the ammonoids known to Mojsisovics, e.g. those from Millibrunnkogel, are from condensed beds and vertical or horizontal fracture fillings, the sort of thing Britons call Neptunian dykes. This means that ammonoids of different ages may occur side by side, or that younger ones may seem to be lower in the sequence than the older. It's not surprising that the old-timers like Suess and Mojsisovics had difficulties in sorting out the sequence and arrived at many false conclusions.

In 1975 Leo went to Timor with Franz Tatzreiter and Edith Kristan-Tollmann. Edith is another friend. In 1966 I was taken to Salzkammergut by Edith and her husband, the well known Vienna professor, Alexander. Alexander had done extensive structural mapping there and knew the way to the Feuerkogel, Sommeraukogel and Steinbergkogel localities. With the assistance of a local farmer he also found Millibrunnkogel. On the path up to the Feuerkogel we rediscovered another old locality – the Ferdinandstollen. It was most interesting. The north side of Feuerkogel was covered with dense pine forest. This was where we should have found Diener's "Mischfauna" (Chapter XII). Only on the second traverse was even a small outcrop found. I confess that Edith and I rested in the col, below, while Alexander went back. She was tired and so was I, but she had better reason. As she later told me, although not realising it at the time, she had only recently conceived their son. Alexander found no more than a piece of limestone with *Halobia*. It is very different now. Leo has opened two quarries and found nearly all of the ammonoids known to Mojsisovics and Diener, and many more besides. He has done the same on the south side, displaying a good section for the Carnian and Lower Norian. His Lower Carnian zonation is particularly welcome because it is the only interval poorly represented in North America. Leo's quarries have been invaded, at intervals, by private collectors. At least one is a friend, who shows Leo the specimens and where they come from. But so far he has kept them. Perhaps Leo, like JP, should engage him in a game of whist, and win them that way ! (Chapter X). Really sad, though, is a story told to me by Leo and Franz. Sommeraukogel, above the town of Hallstatt, is an immensely important locality. Dozens of ammonoids from there were described by Mojsisovics, but not accurately localized. Nearly all were described as being from the "Zone of *Cyrtopleurites bicrenatus*". This zone, as interpreted from this locality gave me great problems in Canada, where, it seemed, at least three zones (Magnus, Rutherfordi, Columbianus) contained Bicrenatus Zone ammonoids. I did not really doubt my own zones, but was influenced to put the Magnus Zone in the Middle Norian, on account of the Sommeraukogel occurrence. Through hard labour Leo and Franz have made great progress in sorting out the Sommeraukogel ammonoid levels. The fossils like those of the Magnus Zone are low in the sequence, as in Canada. I agree that the Magnus Zone may be called Lower, not Middle Norian, but as mentioned in Chapter II, it is breaking the law to call it "Lac". There are still problems at Sommeraukogel and every bit of new information is valuable. This leads to the sad story. Franz and Leo had located a large talus block, preserving ammonoids in several layers, that was sure to contribute important data. Breaking a big block of tough Hallstatt limestone, and recording the exact position of every specimen, is a major operation. When they first discovered the block, they had no time to do this. The job was postponed until the next visit. When they returned, it had been completely demolished by a private collector. Science really lost out that time !

Introducing Edith Kristan-Tollmann, distracted me from describing the Timor Expedition, on which she accompanied Leo Krystyn and Franz Tatzreiter. Coming back to Timor; here Leo and Franz found the blocks described by Burck and others besides (Chapter XI). Lithology, preservation and the ammonoids themselves, are almost exactly as at Sommeraukogel, some 13,000 km away. These they dissected, obtaining ammonoids from up to 10 levels from blocks no more than a metre thick. So far not many of the collections have been described, but Franz has introduced new zones for the late Middle Norian, an important contribution to the time scale.[254]

Leo has also done extensive work on the Upper Trias of Sicily, northern Italy, and Turkey. Particularly important is an occurrence of Hallstatt-type limestone in the Western Taurids of southern Turkey, to which he was introduced by the discoverer, Jean Marcoux of Paris. This provides a record for the Ladinian-Carnian boundary. In 1977, Leo worked out the Jomsom section, north of Annapurna, in Nepal, providing an excellent record for the Upper Carnian and Lower and Middle Norian which agrees closely with the North American sequence.[255] In 1978, with Professor Zapfe , H.M.Kapoor and others of the Geological Survey of India, he performed the ultimate Trias pilgrimage, to Spiti, the place enshrined with memories of the heroical Stoliczka and von Krafft (Chapter IX).[256] Leo and I engage in friendly competition towards discriminating smaller and smaller significant biochronological divisions. Leo has helped me in British Columbia. Twice, in 1980 and 1982, he has joined me and Michael Orchard, and helped to dissect the layers of the Pardonet Formation. Mike collects samples for conodonts from the ammonoid beds in order that the chronology determined from these microfossils can be meshed with that of the ammonoids. In 1982, Franz Tatzreiter came as well. It is exciting work. Leo and Franz could recognize not merely zones but also subzones that they had previously discriminated far away, in Timor and Nepal.

The Himalayan belt has also been studied in recent years by Chinese geologists, notably Wang Yi-gang and He Guo-xiong. In 1966-1968 they had expeditions to Mount Jolmo Lungma and made extensive collections north of the mountain. Jolmo Lungma's more familiar name, to westerners, is Mount Everest.

There have also been notable achievements in the Soviet Union. The discoveries of the last century, on Olenek River and the Sea of Okhotsk, are now realised to have been points on about the biggest area of marine Trias in the world. The Vladivostock area (Primor'ye), another locality discovered early on, has also been extensively restudied.In and bordering Tethys, important faunas have now been described from the Mangyshlak Peninsula and Caucasus. Many Triassic ammonoid faunas are now known. In the first phase of activity the principal worker was Lubov Dmitrievna Kiparisova, a dear lady whom I met in Leningrad in 1966. She died recently, I regret to say, but at a fair age. Yurij Nikolaivitch Popov was more or less her contemporary. He also became a friend but I have not heard from him for the past few years. He worked extensively on faunas from northeastern Siberia. The contemporary younger workers number a dozen or more; Tamara Okuneva, Marianna Victorovna Korchinskya, Tamara Astachova, Sasha Shevyrev, Mischa Vavilov, and Jurij Archipov I had the pleasure to meet in Russia, in 1966. Yurij Zacharov I met when he came to Italy, for the Assereto-Pisa memorial symposium, in 1979. These workers, also Yu. M. Bychkov, A.S. Dagys, S.M.Ermakova and N.K.Zharnikova have made significant contributions to the time scale, particularly for the Lower Triassic and Anisian. For this interval correlation with the North American sequence is pretty clear. The Ladinian and Upper Triassic rocks of northeastern Siberia are thick clastic deposits. Ammonoids are abundant but the faunas lack variety compared with those of North America and Tethys. Occurrences of *Nathorstites* in the Ladinian and of *Eomonotis* and *Monotis* in the Norian facilitate correlation with North America but for the Carnian there are many problems awaiting solution.

Spitsbergen and the other islands of the Svalbard Archipelago, where Trias was discovered in the early days, have received a lot of attention from Norwegian, British, Russian, Polish and German expeditions. Particularly important have been collections and data acquired by Marianna Korchinskya of Leningrad, John Parker, a member of the Cambridge Spitsbergen Expeditions, and, most recently, by Ulrich Lehmann and Wolfgang Weitschat of Hamburg. As now known, the sequence correlates closely with that of North America. It should not be forgotten that the Spitsbergen Trias was studied extensively by Hans Frebold in the 1930s. Hans Wilhelm Ludwig August Hermann Frebold (1899-1983) was born in Hannover. He was educated at the universities in Hannover and Göttingen and in 1931 became a professor at Greifswald. In the 1930s he moved to Copenhagen. Here he established himself as a leading authority on the geology and paleontology of the Arctic. In 1950 he came to Canada where he worked on the Jurassic for the Geological Survey. Official retirement came in 1968, but he continued to work until 1982. Many of the Triassic fossils described by Frebold were collected by other people and did not have accurate data regarding their position. In 1930 Frebold went there as the leader of a Norwegian Expedition. He made collections with accurate data on position and sequence and this led to important discoveries. He was the first to show the position of the latest Triassic *Keyserlingites* fauna (Subrobustus Zone) in relation to the Anisian and Smithian. Circumstances prevented him from fully documenting his discovery. Some 30 years later, when he had moved to Canada to join the Geological Survey, he generously made available what was left of his collection for me to study.[257]

This brings me to the end. I am aware of a glaring omission. Early in the story the Muschelkalk and its *Ceratites* were mentioned. Since then almost nothing has been said. This is because the correlation of the Muschelkalk faunas with those of the Tethys and elsewhere remains one of the biggest problems in Triassic biochronology. Max Urlichs and others are working on it, but there are still few grounds for precise correlation. The interesting Trias of Israel and Spain poses similar problems. Avraham Parnes, who celebrated his 90th birthday in 1983, continues to work on ammonoids from Israel. He took up the study of ammonoids relatively late in his career. When a young man he was a student in Vienna and attended lectures given by Carl Diener (Chapters IX, XII). This presumably stimulated his interest in the Trias. Carmina Virgili has worked on the Spanish fauna. Great progress has been made but there remains a lot to do.

Acknowledgments

For providing copies of the privately printed papers and letters related to the controversy between Bittner and Mojsisovics I am indebted to Dr. Michael Howarth (London), Dr. Ernst Ott (Munich), Dr. Franz Tatzreiter (Vienna), Dr. Gottfried Tichy (Salzburg) and Prof. Dr. H. Zapfe (Vienna). Dr. Tatzreiter also kindly provided the portraits of Arthaber and Bittner. Kathryn Nichols and Norman Silberling (Denver) lent publications from the library of S.W. Muller. George Macdonald Fraser kindly provided information about Yakub Beg. Andre Guex helped with information about the Schlagintweit brothers, his son Jean, with details concerning Scheuchzer's publications.

GENERAL REFERENCES

Alcock, F.J.

1948: A Century in the History of the Geological Survey of Canada; National Museum of Canada, Special contribution 47-1.

Baud, A.

1977: L'échelle stratigraphique du Trias: état des travaux et suggestions; Bulletin du B.R.G.M. (ser.2), sec. 4, no. 3, p. 297-299.

Cohee, G.V., Glaessner, M.F., and Hedberg, H.D. (Eds.)

1978: Contributions to The Geologic Time Scale, Studies in Geology No. 6, American Association of Petroleum Geologists.

Diener, C.

1916: Die Marinen Reiche der Triasperiode; Denkschriften der Akademie der Wissenschaften in Wien, v. 92, p. 405-549.

1925: Leitfossilien der Trias. Wirbellose Tiere und Kalkalgen; Borntraeger, Berlin.

Edwards, W.N.

1967: The Early History of Palaeontology; British Museum (Natural History), Publication 658, London.

Frech, F. (Ed.)

1903-1908: Lethaea geognostica. Handbuch der Erdgeschichte, II, Das Mesozoicum, v. 1, Trias, Stuttgart.

Gaetani, M. (Ed.).

1980: Contributions to the Triassic Stratigraphy, Riccardo Assereto Guilio Pisa Symposium Volume; Rivista Italiana di Paleontologia e stratigrafia, v.85, (3-4).

Harland, W.B., Cox, A.V., Llewellyn, P.G., Pickton, C.A.G., Smith, A.G., and Walters, R.

1982: A geologic time scale; Cambridge University Press.

House, M.R. and Senior, J.R. (Eds.).

1981: The Ammonoidea; Systematics Association Special Volume 18, Academic Press, London and New York.

Hyatt, A. and Smith, J.P.

1905: The Triassic Cephalopod Genera of America; United States Geological Survey, Professional Paper 40.

Kennedy, W.J. and Cobban, W.A.

1976: Aspects of Ammonite biology, biogeography, and biostratigraphy; Special Papers in Palaeontology, no. 17, Palaeontological Association, London.

Kummel, B.

1957: Triassic Ammonoidea, In Treatise on Invertebrate Paleontology, Part L; Geological Society of America and University of Kansas Press.

1979: Triassic, In Treatise on Invertebrate Paleontology, Part A; Geological Society of America and University of Kansas Press.

Lambrecht, K., Quenstedt, W., and Quenstedt, A.

1938: Palaeontologi, Catalogus bio-bibliographicus; Fossilium Catalogus (I), pars 72, Junk Verlag, 's-Gravenhage.

Lehmann, Ulrich.

1981: The ammonites Their life and their world; Cambridge University Press.

Leonardi, P.

1967: Le Dolomiti, Geologia dei Monti tra Isarco e Piave, 2 vols., Trento.

McLearn, F.H.

1953: Correlation of the Triassic Formations of Canada; Bulletin of the Geological Society of America; v. 64, p. 1205-1228.

McLearn, F.H. and Kindle, E.D.
1950: Geology of northeastern British Columbia; Geological Survey of Canada, Memoir 259.

Merrill, G.P.
1924: The First One Hundred Years of American Geology; Yale University Press.

Muller, S.W. and Ferguson, H.G.
1939: Mesozoic Stratigraphy of the Hawthorne and Tonopah Quadrangles, Nevada; Bulletin of the Geological Society of America, v. 50, p. 1573-1624.

Nelson, Clifford M.
1968: Ammonites: Ammon's Horns into Cephalopods; Journal of the Society for the Bibliography of Natural History, v.5, p. 1-18.

Odin, G.S. (Ed.)
1982: Numerical Dating in Stratigraphy; John Wiley and Sons.

Sarjeant, W.A.S.
1980: Geologists and the History of Geology, an International Bibliography from the origins to 1978, 5 vols., Arno Press, New York.

Silberling, N.J.
1959: Pre-Tertiary Stratigraphy and Upper Triassic Paleontology of the Union District Shoshone Mountains Nevada; United States Geological Survey Professional Paper 322.

Silberling, N.J. and Tozer, E.T.
1968: Biostratigraphic classification of the marine Triassic in North America, Geological Society of America, Special Paper 110.

Smith, J.P.
1904: The Comparative Stratigraphy of the Marine Trias of Western America; Proceedings of the California Academy of Sciences, ser. 3, v. 1, no. 10, p. 323-437.

Spath, L.F.
1934: The Ammonoidea of the Trias; Catalogue of the Fossil Cephalopoda in the British Museum (Natural History), Part IV, London.

1951: The Ammonoidea of the Trias (II); Catalogue of the Fossil Cephalopoda in the British Museum (Natural History), Part V, London.

Tozer, E.T.
1958: Stratigraphy of the Lewes River Group (Triassic), Central Laberge Area, Yukon Territory; Geological Survey of Canada, Bulletin 43.
1961: Triassic Stratigraphy and Faunas, Queen Elizabeth Islands, Arctic Archipelago; Geological Survey of Canada, Memoir 316.
1967: A Standard for Triassic Time; Geological Survey of Canada, Bulletin 156.

1982: Marine Triassic Faunas of North America: Their Significance for Assessing Plate and Terrane Movements; Geologische Rundschau, v. 71, p. 1077-1104.

Zapfe, H., (Ed.)
1974: Die Stratigraphie der alpin-mediterranen Trias – The Stratigraphy of the Alpine-Mediterranean Trias; Schriftenreihe Erdwissenschaftlichen Kommissionen, Osterreichische Akademie der Wissenschaften, v. 2.

1978: Beitrage zur Biostratigraphie der Tethys-Trias; Schriftenreihe Erdwissenschaftlichen Kommissionen, Osterreichische Akademie der Wissenschaften, v. 4.

1983: Neue Beitrage zur Biostratigraphie der Tethys Trias; Schriftenreihe Erdwissenschaftlichen Kommissionen, Osterreichische Akademie der Wissenschaften, v. 5.

Zaslow, M.
1975: Reading the Rocks, The story of the Geological Survey of Canada, 1842-1972; Macmillan Company of Canada, Toronto and Department of Energy Mines and Resources, Ottawa.

Zittel, K.A. (transl. M. Ogilvie-Gordon)
1901: History of Geology and Palaeontology to the end of the Nineteenth Century; Walter Scott, London.

Specific References and Notes

Abbreviations: G. B-A., Geologischen Bundesanstalt, Wien; G. R-A., Geologischen Reichsanstalt, Wien.

"op. cit." without designation refers to the General and Appendix references.

1. e.g. the assertion that *Godthaabites*, having a simpler suture line than *Cyclolobus*, is more primitive, and consequently older, despite the fact that the two are nowhere known in sequence (Furnish, W.M., and Glenister, B.F., University of Kansas, Spec. Pub. 4, p. 157 (1970).

2. "The Founders of Geology", The Johns Hopkins Press, Baltimore, p. 98 (1901).

3. Zittel, transl. Ogilvie-Gordon, op. cit. p. 25 (1901).

4. Lambrecht et al., op. cit., p. 256 (1938).

5. Zittel, transl. Ogilvie-Gordon, op. cit., p.460, 466. (1901).

6. Zittel, transl. Ogilvie-Gordon, op. cit., p. 56 (1901).

7. Sir Archibald Geikie, "Life of Sir Roderick I. Murchison", v.1, p. 164, John Murray, London (1875).

8. Zittel, transl. Ogilvie-Gordon, op. cit. p. 63 (1901).

9. J.W.Gregory, Quart. J. Geol.Soc., v. 71, p. liii-lv (1915).

10. Radiometric estimations, 1961-1974, are summarized by R.L. Armstrong, in G.V. Cohee et al., (Ed.) op. cit., p. 90 (1978). More recent estimates are in G.S. Odin and R. Létolle, in G.S. Odin (Ed.) op. cit., p. 527, 532 (1982), and W.B. Harland et al., op. cit. (1982).

11. J.H. Callomon, The measurement of geological time; Jour. Chemical and Physical Soc. University College, London, Spring 1973, p. 39-43 (1973).

The difficulty of relating radiometric ages to the biochronological scale are brought out in Harland et al., op. cit. (1982). They have selected tie points where they have radiometric ages from rocks that they consider to be accurately related to the biochronological scale. For the Triassic they recognize only one, of 218 Ma, which they relate to the Anisian-Ladinian boundary. This is less than satisfactory because they define initial Ladinian in Nevada and use radiometric determinations from localities elsewhere without documenting a biochronological correlation.

12. A clear brief description of the hierarchial system is in J.H. Callomon, and D.T. Donovan, Geol. Mag. v. 103 p. 97, (1966).

13. e.g. by Norman Newell, in Zapfe, H. (Ed.), op. cit., p. 18 (1978).

14. Interpretation of Gangetian as synonymous with Lower Griesbachian was proposed by Tozer in Geol. Surv. Canada, Paper 65-12, p. 7 (1965); Geol. Surv. Canada, Bulletin 156, p. 15 (1967); in H. Zapfe (Ed.), op. cit., p. 198 (1974); and in H. Zapfe (Ed.), op. cit. p. 24 (1978). The alternative interpretation of Gangetian, proposed by L. Krystyn, is in H. Zapfe (Ed.), op. cit., p. 10 (1983).

15. Sitzungberichte Akad. Wiss. Wien, v.104, pp. 1280, 1298 (1895).

16. Abhandlungen G.R-A., v. 6, Supplement, p. 345 (1902).

17. "Upper Triassic ammonoid zones of the Peace River Foothills, British Columbia, and their bearing on the classification of the Norian Stage", Canadian J. Earth Sciences, v.2, p. 216 (1965).

18. See Appendix, p. 144. Details for the North American zones are in N.J. Silberling and K.M. Nichols, "Middle Triassic molluscan fossils of biostratigraphic significance from the Humboldt Range, northwestern Nevada", U.S. Geol. Surv. Prof. Pap. 1207 (1982); N.J. Silberling and E.T. Tozer, op. cit. (1968); E.T. Tozer, op. cit. (1967); E.T. Tozer, "Latest Triassic ammonoid faunas and biochronology, Western Canada", Geol. Surv. Canada Paper 79-1B, pp. 127-135 (1979). The status and significance of the Carnian *Sirenites nanseni* Zone, which appears in earlier tables, is uncertain. This accounts for its absence in Table II. Correlatives of the Desatoyense and Obesum Zones are now recognized in sequence in Pine Pass area, British Columbia. This accounts for their introduction.

For the Tethys the zones listed in Table II are those that can be interpreted from stratotypes. Gaps in the Tethys zonal scheme do not indicate that the Tethys record is incomplete compared with that in North America. These gaps merely indicate intervals for which a well defined zonation is not yet available. A more elaborate summary interpretation of the Tethys zonal scheme has been presented by L. Krystyn, in H. Zapfe (Ed.), op. cit., p.10-11 (1983). Many new sequential data have been acquired in recent years, most of which is provided or summarized in the following publications. *Griesbachian and Nammalian*, K. Nakazawa and H.M.Kapoor (Ed.), "The Upper Permian and Lower Triassic faunas of Kashmir", Pal. Indica, n. ser., v. 46 (1981). *Nammalian and Spathian*, B. Kummel, "Additional Scythian ammonoids from Afghanistan", Bull. Mus. Comp. Zool. Harvard, v. 136, p. 483-508 (1968); Jean Guex, "Le Trias inférieur des Salt Ranges (Pakistan)", Eclogae geol. Helvetiae, v. 71, p. 105-141 (1978). *Spathian-Anisian boundary*, H. Bender, "Der Nachweis von Unter-Anis ("Hydasp") auf der insel Chios", Ann. Geol. Pays Hellen.v.19, p. 412-464 (1970). *Lower-Upper Anisian*, R. Assereto, "Die Binodosus-Zone. Eine Jahrhundert wissenschaftlicher Gegensätze", Sitzungberichte Akad. Wiss. Wien, v. 179, p. 1-29 (1971); R. Assereto, "Aegean and Bithynian: proposal of two new Anisian substages", in H. Zapfe (Ed.), op. cit., p. 23-39 (1974). *Middle Anisian-Lower Carnian*, L. Krystyn and I. Mariolakos, "Stratigraphie und Tektonik der Hallstätterkalk-Scholle von Epidauros (Griechenland), Sitzungberichte Akad. Wiss. Wien, v. 184, p. 181-195 (1975). *Upper Anisian*, H. Rieber, "Ammoniten und Stratigraphie der Grenzbitumenzone (Mittlere Trias) des Monte San Giorgio (Kanton Tessin) Schweiz", Schweiz. Pal. Abhandlungen, v. 73 (1973). *Ladinian*, R. Assereto and O. Monod, "Les formations triasiques du Taurus occidental a Seydisehir (Turquie méridionale)", Riv. Ital. Pal. Strat., Mem. 14, p. 159-191 (1974). *Ladinian-Carnian boundary*, M. Urlichs, "Zur Stratigraphie und Ammonitenfauna der Cassianer Schichten von Cassian (Dolomiten-Italien)", in H. Zapfe (Ed.), op. cit., p. 207-222 (1974); M. Urlichs, "Zur Alterstellung der Pachycardientuffe und der Unteren Cassianer Schichten in den Dolomiten (Italien)", Mitt. Bayer. Staatslg. Hist. Geol., v. 17, p. 15-25 (1977); Edith Kristan-Tollmann and L. Krystyn, "Die Mikrofauna der ladinisch-karnischen Hallstätter Kalke von Saklibeli (Taurus-Gebirge, Türkei) I, Sitzungberichte Akad. Wiss. Wien, v. 184, p. 259-340 (1975). *Lower Carnian*, L. Krystyn, "Eine neue Zonengliederung im alpin-mediterranen Unterkarn", in H. Zapfe (Ed.), op. cit., p. 37-75 (1978). *Carnian-Norian boundary and Lower-Middle Norian*, L. Krystyn, "Stratigraphy of the Hallstatt region", Abhandlungen G.B-A., v. 35, p. 69-98 (1980); L. Krystyn, "Obertriassische Ammonoideen aus dem Zentralnepalesischen Himalaya", Abhandlungen G.B-A., v. 36 (1982). *Middle Norian and Middle-Upper Norian boundary*, F. Tatzreiter, "Ammonitenfauna und Stratigraphie im höheren Nor (Alaun, Trias) der Tethys auf grund neuer Untersuchungen in Timor", Denkschriften Akad. Wiss. Wien, v. 121 (1981).

There are two intervals in Tethys for which adequate sequential data are still not available, namely between the Spathian and Middle Anisian and for the interval of the Amoenum and Crickmayi zones. A zonal and subzonal scheme for the uppermost Triassic has been proposed in J. Wiedmann, F. Fabricius, L. Krystyn, J. Reitner and M. Urlichs, "Über Umfang und Stellung des Rhät", Newsletters Strat., v. 8, p. 145 (1979). This scheme is reproduced by L. Krystyn, in H. Zapfe (Ed.), op. cit, p. 11 (1983). In ascending order the subzones are: 1, *Sagenites quinquepunctatus, 2, Sagenites reticulatus, 3, Vandaites stuerzenbaumi, 4, Choristoceras marshi*. 1 is correlated with the Cordilleranus Zone, 2 with the Amoenum Zone, 3 and 4 with the Crickmayi Zone. This interpretation requires documentation before incorporation into a standard zonal scheme.

Between 1968 and 1979 H. Kozur has published more than 50 papers dealing with Permian and Triassic zonation. Most are listed in, H. Kozur, "Revision der Conodontenzonierung der Mittel- und Obertrias des tethyalen Faunenreichs", Geol. Pal. Mitt. Innsbruck, v.10, (3-4), pp. 79-172 (1980). Kozur chooses to place the Gangetian in the Permian, the Subrobustus Zone in the Anisian, the Sutherlandi Zone in the Carnian, and the Columbianus Zone in the Upper Norian. All his proposals defy recommendations made earlier towards nomenclatural stability, e.g.those in Silberling and Tozer, op. cit. (1968). It seems that Kozur prefers to be different merely for it's own sake.

19. S.M.Ermakova, "Ammonoids and biostratigraphy of the Lower Triassic of Verkhoyansk Range"; U.S.S.R. Academy of Sciences, Siberian Branch, Nauka, Moscow (1981) (in Russian).

20. Namely Kularian, proposed by Y.V. Archipov et. al. in 1971, for parts of the Anisian; Russian and Ussurian (= Ayaxian, = Ajaxian) for the later parts of the Lower Triassic, proposed by Yu. D. Zacharov in 1973 and 1974. Russian is nearly or exactly Spathian; Ussurian etc., Smithian. Detailed discussion and references are given by Tozer in Zapfe, H. (Ed.), op. cit., p. 200 (1974), and in Zapfe, H. (Ed.), op. cit., p. 33 (1978).

21. J. Marwick, "Divisions and faunas of the Hokonui System (Triassic and Jurassic)", N.Z. Geol. Surv., Palaeontological Bull. 21 (1953); Ichikawa, K., "Triassic biochronology of Japan", Proc. 8th Pacific Science Congress, v.2, p. 437 (1956).

22. London and Felling-on-Tyne, The Walter Scott Publishing Company (1910). I have it on the authority of Dr. Hugh Torrens that James Corin was a pen-name of S.S.Buckman.

23. D.J. McLaren, "Time, life and boundaries", J. Paleontology, v. 44, p. 801-815 (1970).

24. A. Hallam, "The End-Triassic Bivalve Extinction Event", Palaeogeography, Palaeoclimatology, Palaeoecology, v. 35, p. 1-44 (1981); R.A.Laws, "Late Triassic depositional environments and molluscan associations from west-central Nevada", ibid., v. 37, p. 131-148 (1982).

25. Suggested correlation between the Manicouagan event and the Triassic-Jurassic boundary was made by D.J. McLaren in "Bolides and biostratigraphy", Bulletin of the Geological Society of America, v. 94, p. 321 (1983). Here will be found reference to the work of the team from the Earth Physics Branch, M.R. Dence, R.A.F. Grieve, and P.B. Robertson, which led to the interpretation of the Manicouagan structure as an impact crater. There is an alternative explanation for the feature; it is shown in the 1:5m Geological Map of Canada (1969) as an area of Triassic volcanic rocks.

26. The question is discussed in more detail by E.T. Tozer in "Triassic Ammonoidea:geographic and stratigraphic distribution", in M.R. House and J.R. Senior (Eds.), op. cit., p. 412-414; and "Latest Triassic (Norian) ammonoid and *Monotis* faunas and correlations", in M. Gaetani (Ed.), op. cit. p. 843-876 (1980). The account of the Rhaetian in W.B. Harland et al., op. cit. (1982) includes misleading inaccuracies. On p. 26 they seem to think that beds with *Rhaetavicula contorta* are younger than Rhaetian. This is nowhere known to be so. On the contrary, according to E. Suess and E. von Mojsisovics (Jahrbuch G.R-A., v. 19, p. 167-200 (1867), the beds with this clam are near the base of the stratotype Rhaetian, below the level of *Choristoceras marshi*. Later on page 26 they state that the type locality for the initial Rhaetian boundary is at Brown Hill, British Columbia. Their statement seems to attribute this designation to me. I have never made this designation, nor has it been made by anybody else. The pages of my own report referred to (Tozer, op. cit., p. 57-60, (1967)) by Harland et al. contain no description of, or reference to Rhaetian. Cordilleranus Zone follows Columbianus Zone at Brown Hill. Possibly Harland et al. propose to regard the base of the Cordilleranus Zone as initial Rhaetian but they make no reference to this zone so their intentions are not clear. On page 114 they indicate that Gümbel regarded the Rhaetian as Lower Jurassic, citing as authoritative J.W. Gregory and B.H. Barrett, "General Stratigraphy", p. 250 (1931). This was a mistake made by Gregory and Barrett. Gümbel regarded the rocks as Triassic.

27. M.J. Fisher, and R.E. Dunay, "Palynology and the Triassic-Jurassic boundary", Review of Palaeobotany and Palynology, v.34 (1981), p. 129-135 (1981); D.G. Smith, Stratigraphic significance of a palynoflora from ammonoid-bearing Early Norian strata in Svalbard, Newsletters Stratigraphy, v.11, p. 154 (1982).

28. Clifford M. Nelson, op. cit. p. 1 (1968).

29. Translated by Mark Chance Bandy and Jean A. Bandy, Geol. Soc. Amer., Special Paper 63 (1955).

30. Zittel, transl. Ogilvie-Gordon, op. cit., p. 16 (1901).

31. W.N. Edwards, op. cit., p. 30 (1967).

32. W.N. Edwards, op. cit., p. 7 (1967).

33. Details on both Scheuchzer and Baier are in Hans Fischer, "Johann Jakob Scheuchzer (2 August 1672 – 23 Juni 1733) Naturforscher und Arzt", Naturforschenden Gesellschaft, Zurich (1973).

34. W.N. Edwards, op. cit., p. 14 (1967).

35. See F.B. Meek, U.S. Geol. Surv. Territories (Hayden Survey), Monograph IX, p.445, (1876).

36. Much of the literature on *Ammonites nodosa* incorporates confusing errors, some of which were introduced when Bruguière provided the original description in "Encyclopédie Méthodique, Histoire Naturelle des Vers," v. 1, p.43, Paris (1792). In referring to the work of Baier, Bruguière cites "Ammonis cornu verrucosa; Bajerus, oryct. norica, pag. 63, tab. 2, fig. 4" as an example of *Ammonites nodosa*. In Baier's work this fossil is named Ammonis cornuum verrucosam and is illustrated by fig. 14, not fig. 4. This error has been repeated in the literature, e.g. by Emil Phillipi, Palaeontologische Abhandlungen, v. 4, p. 410 (1901), and Spath, op. cit., p. 477 (1934). F.A. Quenstedt, "Der Jura", p. 4, Tübingen (1858) gave the correct citation when he asserted that Baier's fossil was an Upper Jurassic ammonite. Bruguière also erred when he stated (loc. cit.) that examples of the ammonoid described and illustrated by Scheuchzer are to be found in the Swiss Mountains. In Scheuchzer's book (p. 259) it is stated explicitly that the fossil was not found in Switzerland. If Bruguière had been right it would be impossible to regard Scheuchzer's fossil as a Muschelkalk *Ceratites*, because there is no true Muschelkalk in the Alps. The title page of Bruguière's book has the date 1792, but according to Henry Dodge, J. Paleont., v. 21, p. 485 (1947), page 43, with the description of *Ammonites nodosa*, was issued in 1789.

37. Wulfen's book is "Abhandlung vom kärnthenschen pfauenschweifigen Helmintholith oder dem sogenannten opalisirenden Muschelmarmor", Erlangen (1793). Sherborn's observation on Wulfen's taxonomy is in "Index Animalium ...", p. vi, Cambridge University Press (1902).

38. "Maris protogaei Nautilos et Argonautas vulgo Cornua Ammonis in Agro Coburgico et vicino reperiundos descripsit et delineavit, simul Observationes de Fossilium Protypis" adjecit D.I.C.M. Reinecke, Cum Tabulis XIII, coloribus expressis, Coburgi,(mdcccxviii). A German translation is now available in Erlanger Geologische Abhandlungen, v. 90 (1972). See also J.F. Pompeckj, "J.C.M. Reinecke, ein deutscher Vorkämpfer der Deszendenzlehre aus dem Anfange des neunzehnten Jahrhunderts", Palaeontologische Zeitschrift, v.8, p. 39 (1927).

39. Specimen Philosophicum Inaugurale, exhibens Monographiam Ammoniteorum et Goniatiteorum, quod, annuente summo numine, ex auctoritate rectoris magnifici Henrici Guilielmi Tydeman, Jc.Ti et antecessoris nec non nobilissimae facultatis philosophicae decreto, pro gradu doctoratus summisque in philosophia honoribus ac privelegiis rite et legitime consequendis defendet Guilielmus de Haan, Amstelodamo-Batavus. Ad diem 7 Maji mdcccxxv hora xi-xii, Lugduni Batavorum, apud H.W.Hazenburg, jun. ii + 168 p.

40. See Leopold von Buch's gesammelte Schriften, 4 vols., Berlin (1867-1885).

41. U. Lehmann, op. cit., p. 219 (1981).

42. C. Diener, Centralblatt, 1908, p. 577 (1908); ibid., 1909, p. 417 (1909); G. Steinmann, ibid. 1909, p. 193 (1909); ibid., 1909, p. 641 (1909).

43. Otto Wilckens, "Gustav Steinmann Sein Leben und Wirken", Geologische Rundschau, v. 21, p. 389-415 (1930).

44. See Arthur Holmes, "Principles of Physical Geology", p. 1126, Nelson, London and Edinburgh (1965).

45. E. Mayr, Evolution, v.6 (4), p. 449 (1952); W.J. Arkell, Systematics Assoc., Publication 2, p. 98 (1956).

46. e.g. Palaeontographica, v.129, (A), p. 135 (1968).

47. Int. Union Geol. Sci., ser. A, v.1, p. 116 (1969).

48. Jean Guex, Bull. Soc. Vaudoise. Sci. Nat., v. 76 (1982), cf. Tozer, in M.R. House and J.R. Senior (Eds.), op. cit. p. 86 (1981).

49. Lambrecht et. al., op. cit., p. 253 (1938).

50. L.F. Spath, op. cit., p. 124 (1951).

51. E.T. Tozer, "Triassic Ammonoidea: Classification, evolution, and relationship with Permian and Jurassic forms", in M.R. House and J.R. Senior, (Eds.), op. cit. p. 65-100 (1981).

52. Walter Carlé, "Friedrich August von Alberti. Schöpfer des Formationsnamens Trias"; Geologische Rundschau, v. 71, p. 705-710 (1982).

53. Zittel transl. Ogilvie-Gordon, op. cit., p. 462 (1901) (Tübingen sandstone should read Täbingen sandstone).

54. Adam Sedgwick and Roderick Impey Murchison, "A Sketch of the Structure of the Eastern Alps; with Sections through the Newer Formations on the Northern Flanks of the Chain, and through the Tertiary Deposits of Styria, etc, etc.", Trans. Geol. Soc. London, ser. 2, v. 3, p. 301-420 (1835).

55. Zittel transl. Ogilvie-Gordon, op. cit., p. 364 (1901).

56. Zittel transl. Ogilvie-Gordon, op. cit., p. 365 (1901).

57. Lambrecht et. al., op. cit., p. 464 (1938).

58. Verhandlungen G. R-A., 1894, p. 184 (1894).

59. A. von Klipstein, "Beiträge zur Geologischen Kenntniss der östlichen Alpen", p. 47, Giessen (1843); "gegen die Verbindung der St. Cassianer Formation mit Muschelkalk" (letter to H. Bronn), Neues Jahrb. p. 800 (1845).

60. Lambrecht et. al., op. cit., p. 128 (1938).

61. "Über die Schichten-folge der Flötz-Gebirge des Gader Thales, der Seisser-Alpe und insbesondere bei St. Cassian", Neues Jahrb., 1844, p. 793 (1844).

62. E. Tietze, "Franz v. Hauer. Sein Lebensgang und seine wissenschaftliche Thätigkeit", Jahrbuch G. R-A., v. 49, p. 679-827 (1900).

63. "Die Geologie und ihre anwendung auf die Kenntniss der Bodenbeschaffenheit der Osterr. – Ungar. Monarchie", Wien, Alfred Holder (1875).

64. Sir Archibald Geikie, "Life of Sir Roderick Impey Murchison", v. 1, p. 166, London, John Murray (1875).

65. Elizabeth Noble Shor, "The Fossil Feud", Exposition Press, Hicksville New York (1974).

66. "Über die Gliederung des Alpen-Kalks in den Ost-Alpen"; Neues Jahrbuch, 1850, p. 584 (1850).

67. Emmrich's interpretation is in "Geognostische Notizen uber den Alpenkalk und seine Gliederung in bairischen Gebirge"; Zeitschrift Deutsch. Geol. Gesell., v. 1, p. 263 (1849). Quenstedt's is in a letter to H. Bronn, Neues Jahrbuch, p. 680-684 (1845) and in "Petrefaktenkunde Deutschlands, Die Cephalopoden", p. 16, Tübingen (1849).

68. "Über die muthmasslichen Äquivalente der Kössener Schichten in Schwaben", Sitzungberichte Akad. Wiss. Wien, v. 21, p. 535-549 (1856).

69. A.E. Nordenskiöld, "The Voyage of the Vega", v. 2, p. 204-209, Macmillan and Co., London, (1881).

70. Lambrecht et al., op. cit., p. 127 (1938).

71. Geographical Journal, v. 3, p. 337 (1894).

72. Quart. J. Geol. Soc., v. 50, p. 53 (1894).

73. Richard A. Pierce, Arctic, v. 35, p. 552 (1982).

74. Sir Clements Markham, "Life of Sir Leopold M'Clintock", p. 52, John Murray, London (1909).

75. Joseph B. MacInnis, "Exploring a 140-year-old ship under Arctic Ice", National Geographic Magazine, v. 164, no. 1, p. 104A (1983).

76. Sir Clements Markham, op. cit. (74), p. 68 (1909).

77. Sir Edward Belcher, "The Last of the Arctic Voyages. . .", v. 2, p. 391, Lovell Reeve, London (1855).

78. Rev. Samuel Haughton, "Geological Account of the Arctic Archipelago, Appendix IV in Sir Leopold M'Clintock, "The Voyage of the Fox. . .", p. 393, John Murray, London (1859).

79. Sir Clements Markham, op. cit (74), p. 187 (1909).

80. Sir Edward Belcher, op. cit. (77), v. 2, p. 47 (1855).

81. W. W. Heywood, "Arctic Piercement Domes". Trans. Can. Inst. Mining and Metallurgy, v. 58, p. 27-32 (1955).

82. F.A. Bather, "Gustav Lindström", Geol. Mag., Decade IV, v. 8, p. 333-336 (1901).

83. Quart. J. Geol. Soc., v. 58, p. lii-liii (1902).

84. A.E. Nordenskiöld, "Sketch of the Geology of Spitzbergen" translated from Trans. Royal Swedish Acad. Sci., p. 27 (1867).

85. H.F. Blanford, J. Asiatic Soc. Bengal, v. 32, p. 124 (1864).

86. Clifford M. Nelson, op. cit., p. 1 (1968).

87. Quart. J. Geol.Soc., v. 49, p. 52-54 (1893).

88. Quart. J. Geol. Soc., v. 62, p.lvi-lx (1906).

89. Quart. J. Geol. Soc., v. 64, p. lix-lxi (1908).

90. Captain Richard Strachey, "On the Geology of Part of the Himalaya Mountains and Tibet", Quart. J. Geol. Soc., v. 7, p. 304 (1851).

91. Quart. J. Geol. Soc., v. 26, p. xxxvi-xxxix (1870).

92. Verhandlungen G.R-A., in Jahrbuch G.R-A., v. 12, p. 258 (1862).

93. Lambrecht et. al., op. cit., p.37 (1938).

94. Sir Roderick Murchison, J. Royal Geog. Soc., v. 27, p. clvi (1857); ibid., v. 28, p. clxxxiii (1858); ibid., v. 29, p. cxxxvii (1859).

95. J. Royal Geog. Soc., v. 40, p. 82 (1870).

96. Sitzungberichte Akad. Wiss. München, 1865, II, p. 385 (1865).

97. Lambrecht et al., op. cit., p. 241 (1938).

98. Geol. Mag., p. 527 (1884).

99. Lambrecht et al., op. cit., p. 309 (1938).

100. E.T. Brewster, "Life and Letters of Josiah Dwight Whitney", Boston, 1909.

101. American Naturalist, v. 12. p. 494 (1878).

102. G.P. Merrill, op. cit., p. 537 (1924).

103. G.P. Merrill, op. cit., p. 523 (1924).

104. G.P. Merrill, op. cit., p. 527 (1924).

105. G.P. Merrill, op. cit., p. 435 (1924).

106. in Alph.- L. Pinart, "Voyages a la Cote Nord-Ouest de L'Amérique exécutés durant les années 1870-72", v. I, partie I, Ernest Leroux, Paris.

107. For details on Selwyn see F.J. Alcock, op. cit., p. 26-40 (1948), and M. Zaslow, op. cit. (1975), p. 131-150; for J. Macoun, "Autobiography", Ottawa Field-Naturalists Club, Special Publication 1 (1979).

108. Lois Winslow-Spragge, "Life and Letters of George Mercer Dawson, 1849-1901", privately printed, Montreal (1962); Joyce C. Barkhouse, "George Dawson, the little giant", Clarke Irwin, Toronto (1974).

109. Trans. Royal Can. Inst., v. 20, p. 1-48 (1934).

110. George Hanson, "Richard George McConnell", Proc. Royal Soc. Canada, 1942, p. 97-98 (1943).

111. R.G. McConnell, "Report on an exploration in the Yukon and Mackenzie Basins, N.W.T."; Ann. Rept. Geological Survey of Canada, 1888-1889, v. 4, part D (1891).

112. For a contemporary account of this stretch of Liard River see Ferdi Wenger, "Canoeing the wild Liard River", Canadian Geographical Journal, v. 93, p. 4-13 (1976).

113. Quart. J. Geol. Soc., v. 66, p. xlix (1910).

114. Centralblatt, 1909, p. 1 (1909); ibid., 1909, p. 616 (1909).

115. Carl Diener wrote a sympathetic biography in 1907; E. Tietze one less so; see Lambrecht et al., op. cit., p. 299 (1938).

116. L. Roth, "Johann Böckh de Nagysur", Verhandlungen G. R-A., 1909 (8), p. 180-181 (1909).

117. "Ueber die Gliederung der oberen Triasbildungen der östlichen Alpen"; Jahrbuch G. R-A., v. 19, p. 127 (1869).

118. F. von Hauer, op. cit. (63), p.334, 351 (1875); Dionys Stur, "Geologie der Steiermark", Tabellarische Uebersicht der Trias-Ablagerungen.., Graz (1871). An account giving the complicated structural interpretation offered nowadays for the area considered by Hauer is in W.Medwenitsch, "Die Geologie der Salzlagerstätten Bad Ischl und Altaussee (Salzkammergut)", Mitteilungen Geol. Ges. Wien, v. 50, p. 133-200 (1957).

119. "Faunengebiete und Faciesgebilde der Trias-Periode in den Ost-Alpen"; Jahrbuch G. R-A., v. 24, p. 84 (1874).

120. Abhandlungen G. R-A., v. 6, (1873-1902).

121. The brief account of Dittmar's career in Annuaire Géologique et Minéralogique de la Russie, v.7, p.38 (1905) makes no reference to the work that he did in western Europe. Dittmar's stratigraphic interpretations for the Hallstatt faunas are in "Zur fauna der Hallstadter Kalke...", Munich (1866). Suess' interpretation of the geology of the Sandling area is in F. von Hauer, "Nachträge zur Kenntniss der Cephalopoden-Fauna der Hallstätter Schichten", Sitzungberichte Akad. Wiss. Wien, v. 41, p. 113-114 (1860).

122. "Die Dolomit-riffe von Südtirol und Venetien", Wien (1879).

123. "Geognostische Beschreibung der Umgegend von Predazzo, Sanct Cassian, und der Seisser Alpe in Süd-Tyrol", Gotha (1860).

124. Manfred Freiherr von Richthofen (1892-1918).

125. E.T. Brewster, op. cit. (100), p. 240 (1909).

126. "Faunistische Ergebnisse aus der Untersuchung der Ammoneen-faunen der Mediterranen Trias", Abhandlungen G. R-A., v. 6, p. 810 (1893).

127. "Entwurf einer Gliederung der pelagischen Sedimente des Trias-Systems"; Sitzungberichte Akad. Wiss. Wien, v. 104, p. 1296 (1895).

128. "Über die Triadischen Pelcypoden-Gattungen *Daonella* und *Halobia*"; Abhandlungen G. R-A., v. 7, (2), p. 37 (1874).

129. "Die Hallstätter Entwicklung der Trias"; Sitzungberichte Akad. Wiss. Wien, v. 101, p. 775-776 (1892).

130. F. Tatzreiter, "Ammonitenfauna und Stratigraphie im höheren Nor (Alaun, Trias) der Tethys aufgrund neuer Untersuchungen in Timor"; Denkschriften Akad. Wiss. Wien, v. 121, p. 65 (1981).

131. Verhandlungen G. R-A., 1902 (6), p. 165 (1902).

132. Jahrbuch G. R-A., v. 42, p. 387 (1893).

133. "Zur neueren Literatur der alpinen Trias"; Jahrbuch G. R-A., v. 44, p. 233 (1895).

134. A. Bittner, "Zur definitiven Feststellung des Begriffes "norisch" in der alpinen Trias", 16 p., Bruder Hollinek, Wien (1895).

135. A. Bittner, "Bemerkungen zur neuesten Nomenclatur der alpinen Trias", 32 p., Bruder Hollinek, Wien (1896).

136. Verhandlungen G. R-A., 1896 (5), p. 191 (1896).

137. E. von Mojsisovics, "Über den chronologischen Umfang des Dachsteinkalkes"; Sitzungberichte Akad. Wiss. Wien, v. 105, p. 5-40 (1896).

138. A. Bittner, "Dachsteinkalk und Hallstätter Kalk. Ein weiterer Beitrag zur Kennzeichnung der "wissenschaftlichen" Thätigkeit des Präsidenten der Trifailer Kohlenwerks-Actien-Gesellschaft, Herrn J.U. Dr. E. v. Mojsisovics, Vicedirectors der k.k. Geolog. Reichsanstalt, wirkl. Mitgliedes der kais. Akademie der Wissenschaften etc. etc."; 80 p., Bruder Hollinek, Wien (1896). "J.U." indicates that Mojsisovics' doctorate was in law.

139. "Herr E.v. Mojsisovics und die öffentliche Moral", 8 p.(no date).

140. "Hochgeehrter Herr College !", letter to "Seiner Hochwohlgeboren Herrn Dr. Edmund v. Mojsisovics", 2 p.,(24 February 1898).

141. "Euer Hochwohlgeboren !", letter to "Herrn Dr. Edmund v. Mojsisovics", 2 p.,(26 February 1898).

142. "Entgegnung auf die Schrift der fünfunddreissig wirklichen Mitglieder der kais. Akademie der Wissenschaften in der Angelegenheit des Herrn. E. v. Mojsisovics", 2 p.,(6 March 1898).

143. E. von Mojsisovics, "Zur Abwehr gegen Herrn Dr. Alexander Bittner", 10 p., Selbstverlag der Verfassers, Wien (1898).

144. 4 p., Verlag von Dr. A. von Böhm, Wien (Ende Marz, 1898).

145. Verhandlungen G. B-A., 1930 (11), p. 231 (1930).

146. "Hochgeehrter Herr !", letter to Dr. August v. Böhm, Graz, 1 p. (28 March 1898).

147. "Briefe zur Nomenclatur der oberen Trias", Druck und Verlag Alfred Holzhausen, Wien (1898). The letter to Mojsisovics, dated 12 April, comprises pages 1-6; Mojsisovics' reply, dated 14 April, continues the pagination (p. 7-9). Presumably the letter and reply were printed and distributed simultaneously.

148. "Recht und Wahrheit in der nomenclatur der oberen alpinen Trias", 31 p., R. Lechner (Wilhelm Müller), Wien (1898).

149. "Eine Bemerkung zur Nomenclatur und Gliederung der alpinen Trias", 6 p., Wien (April 1899).

150. "Die Glaubwürdigkeit des Herrn E. v. Mojsisovics von München aus beleuchtet", 13 p., Wien (November 1899).

151. "Zur Literaturgeschichte der alpinen Trias. Schreiben des Herrn Geheimrathes Prof. Dr. K.A.v.Zittel in München an Herrn Prof. Eduard Suess in Wien", München, 24 December 1899, 4 p. Wien (1899).

152. "Entgegnung auf das von Geheimrath Prof. K.A.v. Zittel in München an Prof. E. Suess in Wien gerichtete Schreiben: "Zur Literaturgeschichte der alpinen Trias'", 10 p. Wien, (8 January 1900), appended is a letter from Dr. A. v. Böhm to Bittner (12 January 1900).

153. "Nachträgliche Bemerkung zu meiner Entgegnung auf das von Herrn Geheimrath Zittel an Herrn Prof. E. Suess gerichtete Schreiben: 'Zur Literaturgeschichte der alpinen Trias' ", 4 p. Wien (15 March 1900). Professor H.Zapfe has Bittner's own copy of this pamphlet. On the last page Bittner had attached a cutting with the announcement of Mojsisovics's retirement as Vicedirector. The cutting is from Verhandlungen G. R-A., 1900 (17-18), p. 387 (1900). Ill health was given as the reason for retirement. Bittner evidently considered that there was a different reason. In his own handwriting the cutting is prefaced with the following words, "Das Schlussresultat der gegen mich am 10 Jänner eingeleiteten und Ende September abgeschlossenen Disciplinar-Untersuchungen:" ("The final result of the disciplinary investigation against me which started on January 10 and was concluded at the end of September:") Bittner has heavily underscored the reference to Mojsisovics' ill health.

154. "Ueber das Alter des Kalkes mit *Asteroconites radiolaris* von Oberseeland in Karnten", Verhandlungen G. R-A., 1902 (2), p.66, 67 (1902).

155. "Eine Bermerkung zur Anwendung des Terminus lacisch", Verhandlungen G. R-A., 1902 (4), p. 120-122 (1902).

156. B. Plöchinger, "Georg Rosenberg 1897-1969", Verhandlungen G. B-A., 1970 (1), p. 1-5 (1970).

157. Georg Rosenberg, "50 Jahre nach Mojsisovics", Mitteilungen Geol. Ges. Wien, v. 50, p. 293-314 (1958).

158. Abhandlungen G. R-A., v. 6 (1) Supplement, p. 345 (1902).

159. "The Trias of the Himalayas"; Mem. Geol. Surv. India, v. 36 (3) (1912).

160. Geol. Mag., Decade IV, v. 7, p. 432 (1900).

161. Quart. J. Geol. Soc., v. 31, p. xlvii-xlix (1875).

162. Dictionary of National Biography, v. 20, p. 33.

163. George Macdonald Fraser, "Flashman at the Charge", Appendix II, p. 292, Pan Books, London (1974); Eugene Schuyler, "Turkistan ..", v. 2, p. 316, Scribner, Armstrong and Co., New York (1877).

164. "Journey from Leh to Yarkand and Kashgar and exploration of the Yarkand River"; Journ. Royal Geographical Soc., v. 40, p. 100 (1870).

165. Geographical Journal, v. 5 (5), p. 409 (1895).

166. Frank E. Younghusband, "The Heart of a Continent", p. 173, John Murray, London (1896).

167. Nature, v. 10, p. 172, 185 (1874); ibid., v. 34, p.574 (1886).

168. Records Geol. Surv. India, v. 8 (1), p. 1 (1875).

169. Records Geol. Surv. India, v. 7 (1), p. 12 (1874); ibid., v. 7 (2), p. 49 (1874).

170. "Geology", Scientific Results of the Second Yarkand Mission; based upon the collections and notes of the late Ferdinand Stoliczka, Ph. D." 49 p., Calcutta (1878).

171. Denkschriften Akad. Wiss. Wien, v. 61, p. 1-38 (1894).

172. P. Martin Duncan, "Karakoram Stones or Syringosphaeridae", Scientific Results of the Second Yarkand Mission; .., 17 p., Calcutta (1879); H.B. Medlicott and W.T. Blanford, "A Manual of the Geology of India, Part II, Extra-Peninsula Area", p. 655, Calcutta (1879). Some of the Karakoram stones were described as a new genus – *Stoliczkaria* – but it is now considered that *Heterastridium* and *Stoliczkaria* are the same (Erik Flügel, J. Paleontology, v. 34, p. 130 (1960).

173. Quart. J. Geol. Soc., v. 64, p. lxiii (1908).

174. R. Etheridge, Trans. Cardiff Nat. Soc., v. 3, p. 39-64 (1872).

175. C. Diener, "Ergebnisse einer geologischen Expedition in den Central- Himalaya von Johar, Hundes und Painkhanda"; Denkschriften Akad. Wiss. Wien, v. 62, p. 533-608 (1895).

176. Alois Kieslinger, "Das Lebenswerk Carl Dieners", Der Geologe, no. 43, p. 1123-1132; ibid, no. 44, p. 1201-1218 (1928).

177. Arnold Heim and August Gansser, "The Throne of the Gods, An Account of the First Swiss Expedition to the Himalayas", Macmillan, New York (1939).

178. J.B. Auden and A.M.N. Ghosh, "Preliminary Account of the Earthquake of 15th January 1934 in Bihar and Nepal", Records Geol. Surv. India, v. 68, p. 177-239 (1934); J.B.Auden, "Traverses in the Himalaya", ibid., v. 69, p. 124-167 (1935). The fossils discovered by Auden in Sikkim were identified and dated by M.R. Sahni, ibid., v. 72, p. 21 (1937).

179. G. Geyer, "Albrecht von Krafft", Verhandlungen G. R-A., 1901 (11-12), p. 261-263 (1901).

180. C. Diener, "Zur Erinnerung an Albrecht von Krafft", Jahrbuch G. R-A., v. 51, p. 154 (1902).

181. F.B. Plummer, "Memorial – James Perrin Smith", J. Paleontology, v. 5, p. 168-170 (1931).

182. W.O. Crosby, "Memoir of Alpheus Hyatt", Bull. Geol. Soc. Amer., v. 14, p. 504-512 (1902).

183. Herbert Clark Hoover and Lou Henry Hoover, "Georgius Agricola, De Re Metallica, translated from the first Latin Edition of 1556. . .", 637 p., The Mining Magazine, London (1912).

184. Eliot Blackwelder, "James Perrin Smith 1864-1931"; Biographical Memoirs, v. 38, U.S.National Academy of Sciences, p. 302 (1965).

185. James Perrin Smith, "Die Jurabildungen des Kahlberges bei Echte"; Jahrbuch Preussischen geologischen Landesanstalt, v. 12, p. 288-356 (1893).

186. Janet Lewis Zullo, "Annie Montague Alexander: Her work in Paleontology"; Journal of the West, v. 8 (2), p. 183-199 (1969).

187. "The Metamorphic Series of Shasta County, California", J. Geology, v. 2, p. 588-612 (1894); "Classification of the Marine Trias", J. Geology, v. 4, p. 385 (1896).

188. Geologie en Mijnbouw, n. ser., 15 (4), p. 104 (1953).

189. F. Tatzreiter, in H. Zapfe (Ed.), op. cit., p. 111 (1978).;

190. Quart. J. Geol. Soc., v. 102, p. xl (1946).

191. Maria M. Ogilvie Gordon, "Das Grödener-, Fassa- und Enneberggebiet in den Südtiroler Dolomiten"; Abhandlungen G. B-A., v. 24 (1), p. ii ("Gedruckt auf Kosten der Verfasserin") (1927).

192. H.P. Cornelius, "Julius von Pia"; Mitteilungen Geologischen Gesellschaft Wien, v. 35, p. 315-324 (1944).

193. J. von Pia, "Maria Mathilda Ogilvie Gordon"; Mitteilungen Geologischen Gesellschaft Wien, v. 32, p. 173-186 (1940).

194. J.F. Pompeckj, "Fritz Frech", Neues Jahrbuch, 1919, p. i-xxxviii (1919).

195. Lambrecht et al., op. cit., p. 336 (1938).

196. Marta Cornelius-Furlani, "Gustav Edler von Arthaber"; Mitteilungen Geologischen Gesellschaft Wien, v. 36-38, p. 297-302 (1949).

197. Lambrecht et al., op. cit., p. 315 (1938).

198. e.g. by H. Kozur, N.J. Newell and J.B. Waterhouse. Newell would place the Griesbachian in the Permian; Waterhouse, Griesbachian and possibly also Dienerian in the Permian. Kozur regards Lower Griesbachian as Permian, correlated with the Dorashamian. Nearly everybody else considers the Dorashamian to be older than Griesbachian. Most authors consider Dorashamian Permian but now B.G.Sokratov (See International Geological Review, v. 25, no. 4, p. 483-496, 1983), although agreeing that Dorashamian is older than Griesbachian prefers to put Dorashamian in the Triassic. References and discussion are given in E.T.Tozer, "The significance of the ammonoids *Paratirolites* and *Otoceras* in correlating the Permian-Triassic boundary beds of Iran and the People's Republic of China"; Can. J. Earth Sciences, v. 16, p. 1524-1532 (1979).

199. "Die Fossillagerstätten in den Hallstätter Kalken des Salzkammergutes; Sitzungberichte Akad. Wiss. Wien, v. 135, p. 73-101 (1926).

200. L.R. Cox, "Leonard Frank Spath", Biographical Memoirs, Fellows of the Royal Society, v. 3, p. 217-226 (1957).

201. F.E. Spath and C.W. Wright, "L.F. Spath, Ammonitologist", Archives Nat. Hist., v. 11, p. 103-105 (1982).

202. Proc. Geologists Ass., v. 70 (1959-1960), p. 123 (1960).

203. Neues Jahrb. Referate III, p. 93, (1935).

204. Norman D. Newell, "Memorial to Bernhard Kummel 1919-1980"; Geological Society of America Memorials, v. 12, 4 p. (1982).

205. W.A.S. Sergeant, op. cit. p. 1744 (1980).

206. B. Kummel, "Triassic Period"; Encyclopedia Britannica, 15th Ed., Macropaedia v. 18, p. 693-697 (1975).

207. E.T. Tozer, "Frank Harris McLearn", Bull. Can. Petroleum Geol., v. 12, p. 932-938 (1964); L.S.Russell, "Frank Harris McLearn", Proc. Royal Soc. Canada, ser. 4, vol. 3, p. 135-139 (1965).

208. Letters, S.W. Muller to F.H. McLearn, January 14, 1936; April 22, 1936.

209. McLearn wrote many papers on this subject. His final conclusions are in "Ammonoid faunas of the Upper Triassic Pardonet Formation, Peace River Foothills, British Columbia", Geological Survey of Canada Memoir 311 (1960). In the discussion given here McLearn's conclusions have been expressed in a simplified terminology, the *Juvavites* level being McLearn's *Pterotoceras – Cyrtopleurites magnificus* bed, and the *Cyrtopleurites* level being the *Drepanites* bed and overlying *Cyrtopleurites* bed. A detailed discussion is in E.T. Tozer, op. cit. (17)(1965).

210. Peter Charlebois, "Sternwheelers and Sidewheelers", NC Press, Toronto, 1978; Norman Soars, "An Era Closes", The Beaver, 1953 (March), p. 28-32 (1953).

211. R.M. Patterson, "Peace River Passage", The Beaver, Winter 1956, p. 14-19.

212. Williams, M.Y. and Bocock, J.B., "Stratigraphy and Palaeontology of the Peace River Valley of British Columbia"; Trans. Royal Soc. Canada, ser. 3, v. 26, sec. 4, p. 197-224 (1932).

213. V.J. Okulitch, "Merton Yarwood Williams"; Proc. Royal Soc. Canada, ser.4, v. 12, p. 87-88 (1975).

214. L.S. Russell, "Charles Mortram Sternberg"; Proc. Royal Soc. Canada, ser.4, v. 20, p.131-135 (1983).

215. Gordon E. Bowes, "Peace River Chronicles", p. 439-457, Prescott Publishing Co., Vancouver (1963); Bruce Ramsey and Dan Murray, "The Big Dam country", Dan Murray Ltd., Fort St. John, British Columbia, 2nd Edition (1972); Bob White, "Bannock and Beans, Memoirs of the Bedaux Expedition and Northern B.C. Packtrails", 371 p., Gateway Publishing Company, Winnipeg (1983).

216. Frances Donaldson, "Edward VIII", p.350, J.B. Lippincott, Philadelphia and New York (1975).

217. J. Bryan III and Charles J.V. Murphy, "The Windsor Story", William Morrow and Co., New York (1979); Diana Mosley, "The Duchess of Windsor", Nelson Canada (1981).

218. S.C. Ells, "Alaska Highway"; Canadian Geographical Journal, v. 28, p. 104-119 (1944).

219. R.W.Boyle, "Edward Darwin Kindle", Geogram, No. 19, Geological Survey of Canada, 1983, p.4 (1983).

220. Alice E. Wilson, "Edward Martin Kindle", Proc. Royal Soc. Canada, 1941, p. 127-130 (1941).

221. Benjamin M. Page, Norman J. Silberling and A. Myra Keen, "Siemon W. Muller", Geological Society of America Memorials, v. 4, p. 142-146 (1975).

222. American J. of Science, v. 244 (9), p. 635 (1946).

223. Bulletin of the Geological Society of America, v. 47, p. 241-252 (1936); ibid., v. 50, p. 1573-1624 (1939).

224. Commemorazione del Prof. Gaetano Giorgio Gemmellaro tenuta nell'Universita di Palermo il 16 Marzo 1905; Annuario della R. Universita di Palermo, Anno scolastico 1905-1906, p. 4-50 (1906).

225. L. Krystyn, "Triassic conodont localities of the Salzkammergut Region"; Abhandlungen G. B-A., v. 35, p. 75 (1980).

226. G.S. Malloch, "Bighorn Coal Basin Alberta"; Geological Survey of Canada Memoir 9-E, p. 43 (1911).

227. The species in question are now placed in *Eoprotrachyceras*.

228. Letter, S.W. Muller to F.H. McLearn, March 4, 1937.

229. Robert R. Shrock, "Memorial to Hervey Woodburn Shimer", Proc. Geological Soc. America (1966), p. 379-385 (1968).

230. J.A. Jeletzky, "Is it possible to quantify biochronological correlation ?", J. Paleontology, v.39, p.135-140 (1965).

231. Olaf Holtedahl, "Summary of Geological Results"; Rept. Second Norwegian Expedition in the "Fram", 1898-1902, no. 36, Kristiania (1917).

232. Canadian Government Hudson Strait and Baffin Island Patrol, 1937, Polar Record, no. 15, p. 38-39 (1938).

233. Markham, op. cit. (74), p. 200 (1909).

234. Y.O. Fortier, "Heavy helicopter reconnaissance in the Arctic Archipelago", Geological Survey of Canada, Bulletin 54, p. 55-60 (1959).

235. Robert L. Bates and Julia A. Jackson (Eds.), American Geological Institute Glossary of Geology (Second Edition), p. 173, 276, 589, 597 (1980).

236. Welland W. Phipps, "By Piper Cub to the extreme north of Canada", Arctic Circular (Ottawa), v. 11, no. 1, p. 7-9 (1958); R. Thorsteinsson and E.T.Tozer, "Geological Investigations in the Parry Islands, 1958", Polar Record. v. 9, no. 62, p. 458-461 (1959).

237. Geological Survey of Canada Memoir 284, p. 284 (1957).

238. N.J. Silberling and R.E. Wallace, U.S. Geol. Surv., Prof. Paper 592 (1969); N.J. Silberling and K.M. Nichols, U.S.Geol. Surv., Prof. Paper 1207 (1982).

239. N.J. Silberling and E.T. Tozer, op. cit. (1968).

240. A.A. Stoyanow, American Men and Women of Science, 12th Ed., p. 6168 (1973).

241. Sheng Jing-zang, Chen Chu-zhen, Wang Yigang, Rui Lin, Liao Zhuo-ting and Jiang Na-yan, "On the *"Otoceras"* beds and the Permian-Triassic boundary in the suburbs of Nanjing", Journal of Stratigraphy, v. 6 (1), p. 1-8 (1982). (in Chinese)

242. Terence Armstrong, "Sir James Wordie", Arctic, v. 15, p. 170-171 (1962).

243. J.M. Wordie, "The Cambridge Expedition to East Greenland in 1926", Geog. J., v. 70, p. 225-265 (1927).

244. Fritz Müller, "Lauge Koch", Arctic, v. 17, p. 290-292 (1964).

245. Lauge Koch, "Geologie von Grönland", Borntraeger, Berlin (1935).

246. O.B. Bøggild, Richard Bøgvad, Karen Callisen, Hans Frebold, Helge Gry, Knud Jessen, Victor Madsen, A. Noe-Nygaard, Christian Poulsen, Alfred Rosenkrantz, Curt Teichert, "Remarks upon Lauge Koch: Geologie von Grönland", Medd. fra Dansk Geologisk Forening., v. 8, p. 497-511 (1935).

247. Lauge Koch, "Polemik eller positivt Forskningsarbejde ? (Polemik oder aufbauende Forschungsarbeit ?)", ibid., v. 9, p. 361-365 (1938).

248. O.B. Bøggild, Richard Bøgvad, Karen Callisen, Helge Gry, Knud Jessen, Victor Madsen, A. Noe-Nygaard, Christian Poulsen, Alfred Rosenkrantz, "Til Retledning (For the Reader's Guidance)", ibid. v. 9, p. 365-369 (1938).

249. Rudolph Trümpy, "Triassic of East Greenland", in Gilbert O. Raasch (Ed.), "Geology of the Arctic", v. 1, p. 248-254, University of Toronto Press (1961).

250. Rudolph Trümpy, "Permian System", Meddeleser om Grønland, v. 168 (3), p. 6-9 (1972); C. Teichert and B. Kummel, "Permian-Triassic boundary in the Kap Stosch area, East Greenland", ibid., v. 197 (5) (1976).

251. G.E.G. Westermann, "Memorial to Otto Schindewolf", Geological Society of America Memorials, v. 3, p. 182-185 (1974).

252. Bibliography on recent work in the Salt Range is in J. Guex, Eclogae geol. Helvetiae, v. 71 (1), p. 123-124 (1978).

253. Lambrecht et al., op. cit., p. 433 (1938).

254. op. cit. (130) (1981).

255. "Obertriassische Ammonoideen aus dem Zentralnepalesischen Himalaya"; Abhandlungen G. B-A., v. 36 (1982).

256. Episodes, Geological Newsletter I.U.G.S., 1979 (1), p. 33 (1979). The unidentified figure in the picture, on the left, is Leopold Krystyn.

257. Frebold's field observations are in "Fazielle Verhaltnisse des Mesozoicums im Eisfjordgebiet Spitzbergens...", Skrifter om Svalbard og Ishavet, v. 37 (1931). This report is titled "I Teil". The second part, which would have included description of the fossils, was never published. The best *Keyserlingites* collected by Dr. Frebold was eventually described in Geological Survey of Canada Bulletin 123, p. 10 (1965).

258. E. Irving, "Drift of the major continental blocks since the Devonian", Nature, v. 270, p. 304-309 (1977). In the Permian and for much of the Trias the equator may have been in a more northerly position by about 20 degrees (See A. Hallam, "Supposed Permo-Triassic megashear between Laurasia and Gondwana" , Nature, v. 301, p. 499-502 (1983).

259. E.T. Tozer, op. cit. (1982).

APPENDIX

Definition and interpretation, Triassic Series, Stages and Substages.

Names used on Table II are in **bold face**.

Aegean Substage. *Paracrochordiceras* beds, Mount Marathovouno, Chios Island, Greece, in the Aegean Sea (Assereto, 1974, p. 36); = **Lower Anisian Substage** .

Ajaxian Stage. *Hedenstroemia bosphorense* and *Anasibirites nevolini* zones, Ajax Bay, Russian Island, Ussuri Bay, U.S.S.R. (Zacharov, 1978, p. 16); = **Smithian Substage** .

Alaunian Substage. Bicrenatus Zone, Sommeraukogel, Austria, named for Alauns, who lived in Hallein area in Roman times (Mojsisovics, 1895, p. 1298; Kuehn, 1962, p. 9); current scope, = **Middle Norian Substage** (Krystyn, 1980, p. 72).

Anatolian Stage. Proposed by R. Assereto in a lecture at Vienna in 1973 for **Aegean** + Bithynian substages; referred to, but no longer formally proposed in Assereto (1974, p. 32, 33); scope = **Caurus** + **Hyatti Zones**; name from Anatolia, the Asian portion of what is now Turkey.

Anisian Stage. Limestone formations at Grossreifling, Enns (=Anisus) River, Austria; conceived as comprising Balatonian and Bosnian substages (Waagen and Diener, 1895, p. 1292; Zapfe, 1971, p. 580; Assereto, 1971; Summesberger and Wagner, 1972); recommended scope, **Caurus Zone** to **Occidentalis Zone** inclusive (Silberling and Tozer, 1968, p.10).There is no historical justification for including the **Subrobustus** and **Haugi** zones in the Anisian, as by Kozur (1980).

Arabbian Substage. Buchenstein Beds (*Protrachyceras curionii* Zone), vicinity of Arabba, Dolomites, northern Italy. Name tentatively suggested by Bittner (1896, p. 12); = early **Ladinian**.

Ayaxian Stage. = Ajaxian **q.v.**.

Badiotic Substage ("Gruppe"). Conceived as early Carnian, heterogeneous, including beds in Austria and Italy now regarded as Ladinian and Carnian; named for people who inhabited the St. Cassian area in Roman times (Mojsisovics, 1869, p. 129, Kuehn, 1962, p. 41).

Bajuvarian Series. (= Juvavian Stage + Rhaetian Stage) (Mojsisovics, 1895, p. 1279).

Balatonian Substage. **Binodosus Zone** (or *Balatonites balatonicus* Zone) in the vicinity of Lake Balaton, Hungary (Waagen and Diener, 1895, p. 1293); = **Pelsonian Substage**.

Balfour Series. Village of Balfour, New Zealand, = Otamitan + Warepan + Otapirian Stages, **q.v.** (Marwick, 1953, p. 15).

Bavarian Stage. Named for Contorta Zone of Bavaria (Slavin, 1961, p. 76, Tozer, 1980, p. 851); ? = **Upper Norian Substage**.

Bithynian Substage. **Osmani** + **Ismidicus Zones**, Kokaeli Peninsula (Bithynia), Turkey (Assereto, 1974, p. 35); = **Hyatti Zone** (Silberling and Nichols, 1982, p. 12). Regarded as inseparable from Pelsonian Substage (Tozer, 1981, p. 409).

Bosnian Substage. Conceived for **Trinodosus Zone** of Reutte and Schreyer Alm in Austria and Han Bulog in Bosnia (Yugoslavia) (Waagen and Diener, 1895, p. 1293) (See Illyrian Stage).

Brahmanian Stage. = **Gangetian** + Gandarian Substages, **q.v.** (Waagen and Diener, 1895, p. 1287).

Cadorian Substage. Marmolada Limestone (Avisianus Zone), Dolomites, northern Italy. Name is from the range which includes the Marmolada summit. Name tentatively suggested by Bittner (1896, p. 13). **Upper Anisian** according to Assereto (1969).

Carnian Stage. Type locality vague, defined to include **Trachyceras** and **Tropites** beds of Hallstatt Limestone, also beds at Raibl, Bleiberg and St Cassian (Mojsisovics 1869, p. 127; Kuehn, 1962, pp. 237-239). Localities are in and around the Austrian state Kärnten (Carinthia). Lieberman (1980) regards beds at Raibl as stratotype. Current scope **Julian** and **Tuvalian Substages** (Krystyn, 1980, p. 72).

Conchylian Stage. Bunter + Muschelkalk, Germany and France (d'Orbigny, 1852, p. 384).

Cordevolian Substage. St. Cassian Beds, Italy, named for the Cordevol people who lived in the type area (Mojsisovics, 1895, p. 1298). Now united with **Julian Substage** (Krystyn, 1980, p. 72).

Dienerian Stage. **Candidus** + **Sverdrupi Zones**, Blind Fiord Formation, Diener Creek, Ellesmere Island (Tozer, 1965, p. 1). Now treated as substage of Nammalian Stage (Tozer, 1981, p. 403).

Dinarian Series. = Hydaspian + **Anisian** Stages, **q.v.** (Waagen and Diener, 1895, p. 1289).

Ellesmerian Substage. = **Upper Griesbachian Substage, q.v.** (Kozur, 1972b, p. 373).

Etalian Stage. Beaumont Stream, New Zealand, named from Mount Etal (Marwick., 1953, p. 12). At least in part **Anisian**.

Fassanian Substage. Buchenstein beds and Marmolada Limestone, northern Italy, name from Val di Fassa (Mojsisovics, 1895, p. 1297); early **Ladinian**.

Franconian Stage. For Muschelkalk of Franconia (Germany) (De Lapparent, 1883, p. 815).

Fujinohiran Stage (Age). Zohoin Group, Sakawa Basin, SW Japan (Ichikawa, 1950, p. 21; 1956, p. 439); part of **Ladinian**.

Gandarian Substage. Lower Ceratite Limestone + Ceratite Marls, Salt Range, Pakistan, named for people who lived in the area in the time of Alexander the Great (Waagen and Diener, 1895, p. 1287). Scope is approximately that of **Dienerian**.

Gangetian Substage. = **Woodwardi Zone** of Himalayas, which occurs near headwaters of the Ganges (Waagen and Diener, 1895, p. 1287); = **Lower Griesbachian Substage** (Tozer, 1965, p. 7).

Gore Series. = Etalian + Kaihikuan + Oretian Stages, **q.v.**; town of Gore, New Zealand (Marwick, 1953, p. 12).

Griesbachian Stage. **Concavum** + **Boreale** + **Commune** + **Strigatus Zones**, Blind Fiord Formation, Griesbach Creek, Axel Heiberg Island (Tozer, 1965, p.1; 1967, p. 13).

Halorian Substage ("Gruppe"). Conceived as late Norian; heterogeneous, named for Celtic people who formerly inhabited Salzkammergut area (Mojsisovics, 1869, p. 128; Kuehn, 1962, p. 197).

Hydaspian Stage. Upper Ceratite Limestone, Salt Range, Pakistan, name is ancient designation for part of Salt Range (Waagen and Diener, 1895, p. 1291); = **Tardus Zone**; incorrectly used for **Lower Anisian** by Pia (1930, p.97).

Illyrian Stage = Bosnian Substage, **q.v.**. Name is translation to Latin, recommended by Pia (1930, p. 97); now ranked as substage of Anisian (Assereto, 1974, p. 32).

Induan Stage. **Woodwardi** Zone of Himalayas defines base, Ceratite Sandstone of Salt Range originally defined top (Kiparisova and Popov, 1956, p. 844). The Ceratite Sandstone was later excluded (Kiparisova and Popov, 1964, p.97).

Isatomean Stage (Age). Kazakoshi and Inai formations, Kitakami area, Honshu, NE Japan (Ichikawa, 1950, p. 21; 1956, p. 440). Scope falls within **Anisian**.

Jakutian Stage. Conceived to include Werfen Formation of the Alps, Ceratite Sandstone of Salt Range, and Subrobustus Beds of Himalayas and supposed correlatives in Siberia, the land of the Yakuts (Waagen and Diener,

1895, p. 1288). Heterogeneous in concept; adoption for late Lower Triassic stage (Krystyn, 1974, p. 44) not recommended.

Julian Substage. Raibl Formation of southern (Julian) Alps (Mojsisovics, 1895, p. 1298). Current scope whole of **Lower Carnian** (Krystyn, 1980, p. 72). Julian ammonoid faunas have more affinity with those of Ladinian than Tuvalian. This provides a case for ranking **Julian** as a stage terminating the **Middle Triassic Series.**

Juvavian Stage (Mojsisovics, 1892, p. 777), = **Norian Stage**. (See Chapters VII, VIII).

Kaihikuan Stage. Kaihiku "Series", South Island, New Zealand (Marwick, 1953, p. 12; Speden and Keyes, 1981, p. 10); ? **Ladinian**.

Karnian Stage = **Carnian Stage q.v.**

Keuperian Stage. Used by some authors (e.g. de Lapparent, 1883, p. 815) for **Upper Triassic Series**.

Kularian Stage. Defined for **Lower Anisian** + **Middle Anisian** correlatives in northeast U.S.S.R.(Archipov, et al., 1971, p. 313). Not used in later publications (e.g. Dagys et al., 1979, p.198).

Labinian Substage. Defined as division of **Norian** Stage, from deposits at Laba River, Caucasus (Slavin, 1961, p. 76); = **Amoenum** and (or) **Crickmayi** zones (Tozer, 1980, p. 851).

Lacian Substage. "Lower Juvavian", *Cladiscites ruber* and *Sagenites giebeli* zones, Hallstatt Limestone, Austria, name from Roman designation for Salzkammergut area (Mojsisovics, 1895, p. 1298). *Discophyllites patens* zone of Hallstatt limestone included later (Mojsisovics, 1902, p. 345). Partly synonymous with Sevatian; inappropriate designation for **Lower Norian** (Tozer, 1974, p. 203). Faunas originally included are mostly or wholly **Upper Norian**. *Discophyllites patens* beds are **Magnus Zone** (Krystyn, 1980, p. 90). Historically better designation for **Upper Norian** than Sevatian (**q.v.**).

Ladinian Stage. Buchenstein and Wengen beds of the Dolomites, Italy, with possible inclusion of St. Cassian beds; named for the Ladin people, who inhabit the region to this day (Bittner, 1892, p. 392). St. Cassian beds generally excluded, although included by Arthaber (1905, p. 272) and Pia (1930, p. 97). Recommended scope, **Subasperum to Sutherlandi Zones** inclusive.

Langobardian Stage. Etymological reconstruction of Longobardian Substage (**q.v.**), presumably because the name is derived from that of the Langobard people (Pia, 1930, p. 97).

Larian Substage ("Gruppe"). Conceived as late Carnian, heterogeneous, included late Triassic beds around Lake Como (Larius), northern Italy. (Mojsisovics, 1869, p. 129; Kuehn, 1962, p. 267).

Longobardian Substage. Wengen Beds of Dolomites, northern Italy (Mojsisovics, 1895, p. 1298), = late **Ladinian**.

Lower Anisian Substage. **Caurus Zone**, Toad Formation, British Columbia (Tozer, 1967, p. 23; Silberling and Tozer, 1968, p. 11)

Lower Carnian Substage. Desatoyense Zone, Augusta Mountain Formation, Nevada + **Obesum Zone**, Ludington Formation, British Columbia (Tozer, 1967, p. 31; Silberling and Tozer, 1968, p.14). The status of the Nanseni Zone, formerly placed between the **Obesum** and **Dilleri** Zones (Tozer, 1967, p.32) is now in doubt.

Lower Griesbachian Substage. Concavum + **Boreale Zones**, Blind Fiord Formation, Griesbach Creek, Axel Heiberg Island (Tozer, 1967, p. 15).

Lower Norian Substage. Kerri + **Dawsoni** + **Magnus Zones**, Pardonet Formation, Brown Hill, British Columbia (Tozer, 1974, p. 203; 1981, p. 402; Krystyn, 1980, p. 72)

Lower Triassic Series. Griesbachian + **Nammalian** + **Spathian Stages. q.v.**, base at **Concavum Zone.**

Malakovian Stage. Proposed for the "beds containing early Triassic marine faunas in New Zealand", name based on Malakoff Hill, southland (Mutch and Waterhouse, 1965, p.1228; Speden and Keyes, 1981, p. 10).

Matsushiman Stage (Age). Rifu Formation, Kitakami area, Honshu, NE. Japan (Ichikawa, 1950, p. 21; 1956, p. 440), **Middle and (or) Upper Anisian**.

Middle Anisian Substage. Hyatti + Shoshonensis Zones, Prida and Favret Formations, Nevada (Silberling and Nichols, 1982, p. 12). The **Varium Zone** has comparable scope; it will probably be replaced by three zones (Tozer, 1971a, p. 1017).

Middle Norian Substage. Rutherfordi + Columbianus Zones, Pardonet Formation, Brown Hill, British Columbia (Tozer, 1981, p. 402).

Middle Triassic Series. Current scope, **Anisian and Ladinian Stages**, basal division, **Caurus Zone**. Inclusion of **Julian Substage** (q.v.) would result in more natural division.

Nammalian Stage. Lower Ceratite Limestone, Ceratite Marls, Ceratite Sandstone, Upper Ceratite Limestone, Nammal, Salt Range, Pakistan (Guex, 1978, p. 115).

Norian Stage. Originally based on rocks containing *"Ammonites metternichi"* in Hallstatt area, Austria, name from Roman Province south of the Danube (Mojsisovics, 1869, p. 127). Beds with *Cyrtopleurites bicrenatus* at Sommeraukogel, Hallstatt, now regarded as stratotype (Zapfe, 1971, p. 584, Krystyn, Schäffer and Schlager, 1971, p. 628). Norian **sensu** Mojsisovics, 1892, p. 780 = Ladinian. Recommended scope, **Kerri Zone to Crickmayi Zone** inclusive (Tozer, 1981, p. 402)

Oenian Substage ("Gruppe"). Conceived as early **Norian**, heterogeneous, name from Inn (Oenus) River (Mojsisovics, 1869, p. 128; Kuehn, 1962, p. 323).

Olenekian Stage. Olenek beds, Siberia. (Kiparisova and Popov, 1956, p. 844). Restricted to correlatives of **Spathian** by Vavilov and Lozovskiy (1970, p. 98). More comprehensive scope (**Smithian + Spathian** adopted by Dagys et al., (1979, Table 10).

Oretian Stage. Oreti "Series", between Dipton and Caroline, New Zealand (Marwick, 1953, p. 13; Speden and Keyes, 1981, p. 10). **Ladinian** and (or) **Carnian**.

Otamitan Stage. Otamita "Series", Otamita stream, Southland, New Zealand (Marwick, 1953, p. 15). Probably **Lower** and (or) **Middle Norian** (Tozer, 1971a, p. 1019; Speden and Keyes, 1981, p. 10).

Otapirian Stage. Otapiri "Series", Otapiri-Taylors Creek watershed, Southland, New Zealand (Marwick, 1953, p.18; Speden and Keyes, 1981, p. 10); **Amoenum** and (or) **Crickmayi** zones (Tozer, 1980, p. 856).

Pelsonian Stage. = Balatonian (**q.v.**), translated into Latin (Pia, 1930, p. 97); now ranked as substage (Assereto, 1974, p. 32).

Poecilian Stage. **Lower Triassic**, as interpreted by Bunter, name attributed to Brongniart by Renevier (1874, p.245).

Raiblian Stage. Raibl beds, northern Italy; attributed to Stoppani, used by Renevier (1874, p. 244), = **Carnian**.

Recubarian Stage. Recoraro Limestone, northern Italy. Tentatively proposed by Bittner (1896, p. 15); = **Anisian Stage**.

Rhaetian Stage. Named by Gümbel (1861, p. 116) for uppermost Triassic rocks in the northern calcareous Alps. The Kössen beds are generally regarded as the stratotype. Scope is much the same as **Upper Norian** (Tozer, 1981, p. 413).

Russian Stage. *Neocolumbites insignis + Subcolumbites multiformis* Zones, Russian Island, near Vladivostock, U.S.S.R. (Zacharov, 1973, p.55); = **Spathian**.

Sakawan Stage (Age). Lower and Middle Kochigatani group, Sakawa Basin, Shikoku, Japan (Ichikawa, 1950, p. 20; 1956, p. 438). Mostly or wholly **Carnian**.

Saliferan Series. Name given to Keuperian by d'Orbigny (1852, p. 384, = **Upper Triassic Series.**

Saragian Stage (Age). Saragai Group, Kitakami massif, Honshu, NE Japan (Ichikawa,

1950, p. 19, 1956, p. 438); **Columbianus** and **Cordilleranus** zones (Tozer, 1980, p. 851).

Schreyeralmian Stage. Hallstatt Limestone at Schreyeralm near Hallstatt, Austria (Bittner, 1896, table only); = **Anisian** and (?) **Ladinian).**

Scythian Series. Conceived as equivalent to Bunter, and thus the whole of the **Lower Triassic Series** (Waagen and Diener, 1895, p. 1278). Named for the Scythians, who inhabited central Asia and southern Russia 800-700 B.C. Definition in terms of a stratotype is ambiguous (Tozer, 1971b, p. 451).

Sevatian Substage. Stratotype is *Pinacoceras metternichi* Zone and *Sirenites argonautae* Zone, both in Hallstatt area; named for Celtic people who lived between the Inn and Enns Rivers (Mojsisovics, 1895, p. 1298). The beds with *Sirenites argonautae* are now known to be **Middle Norian** (Tozer, 1971a, p. 1020; Tatzreiter, 1981). Kozur (1980) consequently includes **Middle Norian** beds in the Sevatian, drawing the base at the level of the **Columbianus Zone.** Krystyn (1980) draws the Sevatian base at the level of the **Cordilleranus Zone**.

Smithian Stage. **Romunderi** + **Tardus Zones**, Blind Fiord Formation, Smith Creek, Ellesmere Island (Tozer, 1965, p. 1; 1967, p.19); now ranked as substage.

Somersetian Stage. Cotham + Langport + Watchet beds, i.e. strata between Westbury Beds (with *Rhaetavicula contorta*) and Lias, Somerset, England; = late Rhaetian. Name was proposed orally but published only in the account of the discussion "in deference to the wishes of the Publication Committee". (Richardson, 1911, p. 73).

Spathian Stage. **Pilaticus** + **Subrobustus Zones**, Blaa Mountain Formation, Spath Creek, Ellesmere Island (Tozer, 1965, p. 4; 1967, p.20).

Swabian Stage. Erroneus citation of Bavarian Stage (**q.v.**) (Pearson, 1970, p. 145).

Tatean Stage (Age). Hiraiso Formation of Inai Group, Kitakami Massif, Honshu, NE Japan (Ichikawa, 1950, p. 22; 1956, p. 441). **Lower Triassic** but exact age and relationship with other Japanese stages uncertain.

Tirolian Series. "Norian" (i.e. Ladinian) + Carnian Stages (Mojsisovics, 1895, p. 1296).

Tozerian Substage (Kozur, 1972a, p. 19) = **Ellesmerian Substage** q.v.

Tsuyan Stage (Age). Oosawa Formation of Inai group, Kitakami Massif, Honshu, NE Japan (Ichikawa, 1950, p. 21; 1956, p. 440); possibly **Spathian**.

Tuvalian Substage. Conceived for **Subbullatus** Zone, which was known from the Tuval mountains, the Roman name for the area between Berchtesgaden and Hallein (Mojsisovics, 1895, p. 1298). Krystyn and Schlager (1971, p. 604) recommend that the north side of the Feuerkogel, in Salzkammergut, Austria, be adopted as the stratotype. Current scope, whole of the **Upper Carnian** (Krystyn, 1980, p. 72). Tuvalian ammonoids being very different from those of the Julian, a case can be made for ranking Tuvalian as a stage initiating the **Upper Triassic Series q.v.**.

Uonashian Stage (Age). Taho Formation of Uonashi District, Shikoku, SW Japan (Ichikawa, 1950, p. 21; 1956, p. 441), = **Tardus Zone**.

Upper Anisian Substage. Rotelliformis + **Meeki** + **Occidentalis Zones**, Prida Formation, Nevada (Silberling and Tozer, 1968, p.11; Silberling and Nichols, 1982, p. 12).

Upper Carnian Substage. Dilleri + **Welleri** + **Macrolobatus Zones**, Hosselkus Formation, California, Luning Formation Nevada (Silberling and Tozer, 1968, p.14)

Upper Griesbachian Substage. Commune + **Strigatus Zones**, Blind Fiord Formation, Griesbach Creek, Axel Heiberg Island (Tozer, 1967, p.16).

Upper Norian Substage. Cordilleranus + **Amoenum** + **Crickmayi Zones**, Pardonet Formation, Tyaughton Group, British Columbia (Tozer, 1979, p. 128).

Upper Triassic Series. Current scope, **Carnian** and **Norian** Stages, base at **Desatoyense Zone**. Exclusion of **Julian Substage** and ranking **Tuvalian** as a basal Upper Triassic Stage, would be an improvement. With this arrangement basal division of **Upper Triassic Series** would be **Dilleri Zone**.

Ussurian Stage (Zacharov, 1973, p.55), replaced by Ajaxian Stage (Zacharov, 1978, p. 16) **q.v.**.

Vaslenian Stage, *Voltzia* beds, Wasselonne, NE France (DuBois, 1948; see Visscher, 1971, p. 74). Early Lower Triassic.

Verkhoyanian Stage. Correlatives of **Smithian**, Verkhoyansk area, NE Siberia (Vavilov and Lozovskiy, 1970, p. 98).

Vetlugian Stage. Basal Triassic, Vetluga River (tributary of Volga) (Mazarovich, 1934; see Visscher, 1971, p. 74).

Virglorian Stage, Virgloria Limestone, Virgloria Pass, Vorarlberg, Austria, (Renevier, 1874; Kuehn, 1962, p.500) = **Anisian**.

Vogesian (sic) Series or Stage. Named for the Vosges, attributed to Karl Mayer by Renevier (1874); = **Lower Triassic**.

Warepan Stage. Warepa "Series", Warepa District, Otago, New Zealand (Marwick, 1953, p. 17) = **Cordilleranus Zone**.

Werfenian Stage. Werfen Formation, Werfen, Austria (Renevier, 1874; Leonardi, 1967, p. 107;), = **Lower Triassic Series**.

Würtzburgian Series or Stage. Named for Wurtzburg, Germany, attributed to Karl Mayer by Renevier (1874), = Muschelkalk, **Middle Triassic Series**.

APPENDIX REFERENCES

Arkhipov, Yu. V., Bychkov, Yu.M., and Polubotko, I.V.

1971: A new zonal scheme for Triassic deposits from northeast U.S.S.R.; Bulletin Canadian Petroleum Geology, v. 19, p. 313-314.

Arthaber, G. von

1905: Die alpine Trias des Mediterran-Gebiets; Lethaea geognostica, Trias (3). Stuttgart.

Assereto, R.

1969: Sul significato stratigrafico della "Zona ad Avisianus" del Trias medio delle Alpi; Bollettino Societa Geologica Italiana, v. 88, p. 123-145.

1971: Die Binodosus-Zone. Ein Jahrhundert wissenschaftlicher Gegensätze; Sitzungberichte Akademie Wissenschaften in Wien, v. 178, p. 1-29.

1974: Aegean and Bithynian: Proposal for Two New Anisian Substages; Schriftenreihe der Erdwissenschaftlichen Kommissionen, Österreichische Akademie der Wissenschaften, v. 2, p. 23-39.

Bittner, A.

1892: Was ist norisch?; Jahrbuch Geologischen Reichsanstalt, v. 42, p. 387-396.

1896: Bemerkungen zur neuesten Nomenclatur der alpinen Trias; 32p., privately published, Vienna.

Dagys, A.S., Arkhipov, Yu. V., and Bychkov, Yu.M.

1979: Stratigraphy of the Triassic system of North-Eastern Asia; Academy of Sciences, U.S.S.R, Transactions of the Institute of Geology and Geophysics, v. 447 (in Russian).

Gümbel, W.C.

1861: Geognostische Beschreibung des bayerischen Alpengebirges und seines Vorlandes; Gotha.

Guex, J.
1978: Le Trias inférieur des Salt Ranges (Pakistan): problèmes biochronologiques; Eclogae geologicae Helvetiae, v. 71, p. 105-141.

Ichikawa, K.
1950: Chronological classification of the Triassic period in Japan; Journal Geological Society of Japan, v. 56, p. 17-22 (in Japanese).

1956: Triassic biochronology of Japan; Proceedings Eighth Pacific Science Congress, v. II, p. 437-442.

Kiparisova, L.D. and Popov, Yu. N.
1956: Subdivision of the Lower Series of the Triassic System into Stages; Doklady Academy Sciences U.S.S.R., v. 109, p. 842-845 (in Russian)

1964: The project of subdivision of the Lower Triassic into Stages; XXII International Geological Congress, Reports of Soviet geologists, p. 91-99 (in Russian).

Kozur, H.
1972a: Probleme der Triasgliederung und Parallelisierung germanische-tethyale Trias; Symposium Mikrofazies und Mikrofauna der Alpinen Trias und deren Nachbargebiete, Innsbruck, Kurzfassung der Vortrage, p. 18-22.

1972b: Vorläufige Mitteilung zur Parallelisierung der germanischen und tethyalen Trias sowie einige Bemerkungen zur Stufen- und Unterstufengliederung der Trias; Mitteilungen der Gesellschaft der Geologie- und Bergbaustudenten in Österreich, Innsbruck, v. 21, p. 361-412.

1980: Revision der Conodontzonierung der Mittel- und Obertrias des tethyalen Faunenreichs; Geologische Paläontologische Mitteilungen Innsbruck, v. 10 (3-4), p. 79-172.

Krystyn, L.
1974: Die Tirolites-Fauna der untertriassischen Werfener Schichten Europas und ihre stratigraphische Bedeutung; Sitzungberichte Osterreichische Akademie der Wissenschaften, v. 183, p. 29-50.

1978: Eine neue Zonengliederung im alpin-mediterranen Unterkarn; Schriftenreihe Erdwissenschaftlichen Kommissionen Osterreichische Akademie der Wissenschaften, v. 4, p. 37-75.

1980: Stratigraphy of the Hallstatt region; Abhandlungen Geologischen Bundesanstalt Wien, v. 35, p. 69-98.

Krystyn, L., Schäffer, G., and Schlager, W.
1971: Der stratotypus des Nor; Annales Instituti Geologici Publici Hungarici, v. 54, p. 607-629.

Krystyn, L., and Schlager, W.
1971: Der stratotypus des Tuval; Annales Instituti Geologici Publici Hungarici, v. 54, p. 591-605.

Kuehn, O.
1962: Autriche, Lexique Stratigraphique International, v. I, fasc. 8.

de Lapparent, A.
1883: Traité de Géologie, Paris, F. Savy.

Leonardi, P.
1967: Le Dolomiti, Geologi dei Monti tra Isarco e Piave, 2 vols., Trento.

Lieberman, H.M.
1980: The suitability of the Raibl sequence as a stratotype for the Carnian Stage and the Julian Substage of the Triassic; Newsletters Stratigraphy, v. 9 (1), p. 35-42.

Marwick, J.
1953: Divisions and Faunas of the Hokonui System (Triassic and Jurassic); New Zealand Geological Survey Palaeontological Bulletin 21.

Mojsisovics, E. von
1869: Über die Gliederung der oberen Triasbildungen der östlichen Alpen; Jahrbuch Geologischen Reichsanstalt, v. 19, p. 91-150.

1892: Die Hallstätter Entwicklung der Trias; Sitzungberichte Akademie Wissenschaften Wien, v. 101, p. 769-780.

1895: Obere Trias, in Mojsisovics, Waagen and Diener, 1895, p. 1296-1302.

1902: Die Cephalopoden der Hallstätter Kalke; Abhandlungen Geologishen Reichsanstalt, v. 6, part 1 (Supplement).

Mojsisovics, E. von, Waagen, W., and Diener, C.
1895: Entwurf einer Gliederung der pelagischen Sedimente des Trias-Systems; Sitzungberichte Akademie Wissenschaften Wien, v. 104, p. 1271-1302.

Mutch, A.R. and Waterhouse, J.B.
1965: A new local stage for the Early Triassic rocks of New Zealand; New Zealand Journal of Geology and Geophysics, v. 8, p. 1228-1229.

d'Orbigny, A.
1852: Cours élémentaire de Paléontologie et de Géologie stratigraphiques, v. 2, Victor Masson, Paris.

Pearson, D.A.B.
1970: Problems of Rhaetian stratigraphy with special reference to the lower boundary of the stage; Quarterly Journal Geological Society of London, v. 126, p. 125-150.

Pia, J.
1930: Grundbegriffe der Stratigraphie mit ausführlicher anwendung auf die Europäische Mitteltrias; Deuticke, Leipzig und Wien.

Renevier, E.
1874: Tableau des Terrains Sédimentaires qui représentent les Epoques de la Phase organique; Bulletin Société Vaudoise des Sciences Naturelles, v. 13, p. 218-252.

Richardson, L.
1911: The Rhaetic and Contiguous Deposits of West, Mid and part of East Somerset; Quarterly Journal Geological Society of London, v. 67, p. 1-74.

Silberling, N.J. and Nichols, K.M.
1982: Middle Triasic Molluscan Fossils of Biostratigraphic Significance from the Humboldt Range, Northwestern Nevada; United States Geological Survey Professional Paper 1207.

Silberling, N.J. and Tozer, E.T.
1968: Biostratigraphic Classification of the Marine Triassic in North America; Geological Society of America Special Paper 110.

Slavin, V.I.
1961: Stratigraphic position of the Rhaetian Stage; Soviet Geology, 1961 (3), p. 69-78 (in Russian).

Speden, I.G. and Keyes, I.W.
1981: Illustrations of New Zealand Fossils; New Zealand Department of Scientific and Industrial Research, Information Series No. 150.

Summesberger, H. and Wagner, L.
1972: Der Stratotypus des Anis (Trias); Annalen Naturhistorisches Museum Wien, v. 76, p. 515-538.

Tatzreiter, F.
1981: Ammonitenfauna und Stratigraphie im höheren Nor (Alaun,Trias) der Tethys aufgrund neuer Untersuchungen in Timor; Denkschriften Österreichische Akademie der Wissenschaften, v. 121, p. 1-142.

Tozer, E.T.
1965: Lower Triassic Stages and Ammonoid Zones of Arctic Canada; Geological Survey of Canada Paper 65-12.

1967: A Standard for Triassic Time; Geological Survey of Canada Bulletin 156.

1971a: Triassic Time and Ammonoids: Problems and Proposals; Canadian Journal of Earth Sciences, v. 8, p. 989-1031.

1971b: Permian Triassic Boundary in West Pakistan; Geological Magazine, v. 108, p. 451-455.

1974: Definitions and Limits of Triassic Stages and Substages: Suggestions prompted by comparisons between North America and the Alpine-Mediterranean region; Schriftenreihe Erdwissenschaftlichen Kommissionen Österreichische Akademie der Wissenschaften, v. 2, p. 195-206.

1978: Review of the Lower Triassic Ammonoid succession and its bearing on chronostratographic nomenclature; Schriftenreihe Erdwissenschaftlichen Kommissionen Österreichische Akademie der Wissenschaften, v. 4, p. 21-36.

1979: Latest Triassic Ammonoid faunas and biochronology, Western Canada; Geological Survey of Canada Paper 79-1B, p. 127-135.

1980: Latest Triassic (Upper Norian) ammonoid and *Monotis* faunas and correlations; Rivista Italiana di Paleontologia e Stratigrafia, v. 85, p. 843-876.

1981: Triassic Ammonoidea: Geographic and Stratigraphic Distribution; in M.R.House and J.R.Senior (Eds.), The Ammonoidea, Systematics Association Special Volume 18, p. 397-431, Academic Press, London and New York.

Vavilov, M.N. and Lozovskiy, V.R.
1970: To the problem of Lower Triassic stage differentiation; Proceedings U.S.S.R. Academy of Sciences, Geological Series, 1970 (9), p. 93-99 (in Russian).

Visscher, H.
1971: The Permian and Triassic of the Kingscourt outlier, Ireland; Geological Survey of Ireland Special Paper 1.

Waagen, W. and Diener, C.
1895: Untere Trias, in Mojsisovics, Waagen and Diener, 1895, p. 1278-1296.

Zacharov, Yu. D.
1973: New subdivision for stages and zones of Lower division of the Triassic; Academy of Sciences U.S.S.R., Siberian Branch, Geology and Geophysics, 1973 (7), p. 51-58 (in Russian).

1978: Lower Triassic ammonoids of East U.S.S.R., Academy of Sciences U.S.S.R., Far Eastern Scientific Centre, Institute of Biology and Soil, Nauka, Moscow (in Russian).

Zapfe, H.
1971: Die stratotypen des Anis, Tuval und Nor und ihre bedeutung für die biostratigraphie und biostratinomie der Alpinen Trias; Annales Instituti Geologici Publici Hungarici, v. 54, p. 579-590.

CANADIAN TRIASSIC AMMONOIDS (p. 24)

1. *Otoceras concavum* Tozer (18881), Concavum Zone; 2. *Otoceras boreale* Spath (14020),Boreale Zone; 3. *Ophiceras commune* Spath (14030), Commune Zone; 4. *Proptychites strigatus* Tozer (72264), Strigatus Zone; 5. *Proptychites candidus* Tozer (14285), Candidus Zone; 6. *Prionolobus* n.sp. (21768), Candidus Zone; 7. *Vavilovites sverdrupi* Tozer (28096), Sverdrupi Zone; 8. *Kashmirites borealis* Tozer (14073), Romunderi Zone; 9. *Euflemingites romunderi* Tozer (14050), Romunderi Zone; 10. *Juvenites needhami* Tozer (14289), Romunderi Zone; 11. *Paranannites spathi* (Frebold) (14085), Romunderi Zone; 12. *Xenoceltites subevolutus* Spath (28053), Tardus Zone; 13. *Wasatchites canadensis* McLearn (9472), Tardus Zone; 14. *Kashmirites warreni* McLearn (9600), Tardus Zone; 15. *"Olenikites pilaticus"* Tozer (18894), Pilaticus Zone; 16. *Monacanthites monoceros* Tozer (18838), Subrobustus Zone; 17. *Olenikites canadensis* Tozer (14094), Subrobustus Zone; 18. *Lenotropites* n.sp., (28475), Caurus Zone; 19. *Parapinacoceras hagei* (McLearn) (28389), Varium Zone; 20. *Czekanowskites* n.sp. (28428), Varium Zone; 21. *Anagymnotoceras varium* (McLearn) (14233), Varium Zone; 22. *Anagymnotoceras columbianum* (McLearn) (28303), Varium Zone; 23. *Amphipopanoceras tetsa* (McLearn) (28270), Deleeni Zone; 24. *Frechites liardensis* (McLearn) (28327), Deleeni Zone; 25. *Tropigymnites* n.sp. (28375), Deleeni Zone; 26. *Intornites canadensis* (McLearn) (6450), Deleeni Zone; 27. *Tozerites* n.sp.(28564), Chischa Zone; 28. *Frechites* n.sp. (28357), Subasperum Zone; 29. *Progonoceratites poseidon* Tozer (18887), Poseidon Zone; 30. *Thanamites parvus* (McLearn) (9527), Meginae Zone; 31. *Protrachyceras sikanianum* McLearn (9044), Meginae Zone; 32. *Drumoceras* n.sp. (28627), Meginae Zone; 33. *Meginoceras meginae* McLearn (9531), Meginae Zone; 34. *Anolcites* n.sp. (28768),

Maclearni Zone; 35. *Frankites sutherlandi* (McLearn) (28801), Sutherlandi Zone; 36. *Lobites ellipticus* (Hauer) (28951), Sutherlandi Zone; 37. *Nathorstites macconnelli* (Whiteaves) (28015), Sutherlandi Zone; 38. *Daxatina canadensis* (Whiteaves) (4718), Sutherlandi Zone; 39. *Trachyceras desatoyense* Johnston (28823), Desatoyense Zone; 40. *Sirenites nanseni* Tozer (14154), Carnian; 41. *Sympolycyclus* n.sp. (14236), Dilleri Zone; 42. *Trachysagenites erinaceus* (Dittmar) (23066), Welleri Zone; 43. *Tardeceras parvum* Hyatt and Smith (32124), Welleri Zone; 44. *Goniojuvavites kellyi* Smith (14244), Welleri Zone; 45. *Jovites borealis* Tozer (14112), Welleri Zone; 46. *Tropites* n.sp. (18910), Welleri Zone; 47. *Jovites richardsi* Tozer (14188), Welleri Zone; 48. *Homerites semiglobosus* (Hauer) (14235), Welleri Zone; 49. *Discotropites smithi* Kutassy (28996), Welleri Zone; 50. *Anatropites* n.sp. (32108), Macrolobatus Zone; 51. *Stikinoceras kerri* McLearn (22736), Kerri Zone; 52. *Thisbites dawsoni* McLearn (34606), Kerri Zone; 53. *Malayites dawsoni* McLearn (72263), Dawsoni Zone; 54. *Pseudocardioceras acutum* (Mojsisovics) (32234), Dawsoni Zone; 55. *Juvavites magnus* McLearn (8837), Magnus Zone; 56. *Drepanites rutherfordi* Mclearn (28853), Rutherfordi Zone; 57. *Cyrtopleurites bicrenatus* (Hauer) (12576), Rutherfordi Zone; 58. *Parathisbites oineus* McLearn (32309), Columbianus Zone; 59. *Episculites* n.sp.(12600), Columbianus Zone; 60. *Pseudosirenites pardoneti* (McLearn) (28753), Columbianus Zone; 61. *Distichites* n.sp. (28918), Columbianus Zone; 62. *"Thetidites"* n.sp. (28885), Columbianus Zone; 63. *Hypisculites* n.sp. (32123), Columbianus Zone; 64. *Pleurodistichites stotti* Tozer (28921), Columbianus Zone; 65. *Himavatites columbianus* McLearn (9265), Columbianus Zone; 66. *Leislingites* n.sp. (28891), Columbianus Zone; 67. *Tragorhacoceras occultum* (Mojsisovics) (32356), Cordilleranus Zone; 68. *Paraguembelites ludingtoni* Tozer (32281), Cordilleranus Zone; 69. *Peripleurites peruvianum* Wiedmann (32320), Cordilleranus Zone; 70. *Rhabdoceras suessi* Hauer (32311), Cordilleranus Zone; 71. *Lissonites canadensis* Tozer (28932), Cordilleranus Zone; 72. *Choristoceras suttonense* Clapp and Shimer (32322, 32325, 32326), Crickmayi Zone; 73. *Cochloceras canaliculatum* Hauer (32329), Amoenum Zone; 74. *Choristoceras crickmayi* Tozer (18912), Crickmayi Zone.

Index

Bold face, definitive. *Italic*, illustrations, map or table.

Notes added in proof stage.

A comprehensive review of the Triassic radiometric time scale, overlooked on page 9, has been given by John A. Webb, "A radiometric time scale of the Triassic", Journal of the Geological Society of Australia, v. 28, p. 107-121 (1981). The numbers he gives are: Triassic-Jurassic boundary, 200 Ma; Norian-Carnian, 215 Ma; Carnian-Ladinian, 225 Ma; Ladinian-Anisian, 235 Ma; Anisian-Lower Triassic, 240 Ma; Triassic-Permian, 245 Ma. He assigns an error of $\pm$ 5 Ma and provides the quotation: "These figures . . . like those in railway timetables, are subject to change without notice".

Another tabulation, with roughly the same numbers, has been made in connection with the preparation of the 27 synthesis volumes of "The Geology of North America" (Allison R. Palmer, Geology, v. 11, p. 503-504, 1983). In this, in accordance with the recommendation on page 14, the Rhaetian Stage has been eliminated, and the youngest Triassic rocks are included in the Norian.

Photograph Index

Geological Survey of Canada photograph numbers. For each page sequence is left to right, top to bottom.

Page 24 202811
Page 26 204029-I, 201762-S, 204029-U, 201762-R
Page 32 203192-I
Page 42 204029-J, 201762-L
Page 44 201762-N, 204029-R
Page 47 68774, 200022, 68776, 200096
Page 50 95433, 200939-C
Page 51 200939-A, 200939-B
Page 54 204029-N, 204029-N, 201762-Q, 204029-V
Page 56 201762-V
Page 58 201762-U
Page 62 204029-Z, 204029-O, 204029-X, 204029-Q, 204029-W
Page 74 201762-M, 204029-T, 204029-S
Page 86 204029-L, 112055, 204029-P, 204029-K
Page 90 96665, 203192-A
Page 91 85390, 190362
Page 92 122595, 85370
Page 94 203192-E, 203192-F
Page 96 112117-T, 204029-E
Page 98 52275, 200939-D, 122449, 201762-I, 203192-J
Page 104 122255, 122256, 106438
Page 110 110788, 113083-B, 110789
Page 111 109203, 113083-K
Page 112 110343, 202651
Page 114 122662, 122663
Page 116 122671, 122465
Page 118 122288, 122535, 122526
Page 120 113082-N, 201392-D, 201615-A, 201615
Page 122 201762-J, 201762-W